Exporting Communication Technology To Developing Countries

Exporting Communication Technology To Developing Countries

Sociocultural, Economic, and Educational Factors

Emmanuel K. Ngwainmbi

University Press of America,® Inc.
Lanham • New York • Oxford

University Press of America,® Inc.
4720 Boston Way
Lanham, Maryland 20706

12 Hid's Copse Rd.
Cumnor Hill, Oxford OX2 9JJ

Printed in the United States of America
British Library Cataloging in Publication Information Available

ISBN 0-7618-1418-3 (cloth: alk. ppr.)
ISBN 0-7618-1419-1 (pbk: alk. ppr.)

™ The paper used in this publication meets the minimum requirements of American National Standard for Information Sciences—Permanence of Paper for Printed Library Materials, ANSI Z39.48—1984

for my family

Contents

List of Acronyms xii
List of Tables, Charts, and Figures xv
Introduction 1

Part I
Instuctional/Academic Overview of Infotech

Chapter 1 Communication Technology 11

Telecommunication versus Communication Technology 13
Communication Technology 15
Computers as Communication Technology 18
Computer Technology and Development in Africa 33
Communication Tech.: The International Dimension 34
A Brief History of the Information Society 36
Exporting Technology 38
Technology, Government and Law 40

Part II
Infotech and Socioeconomic Development: A Market Overview

Chapter 2 Foreign Telecommunication Companies: Market Dynamics and Services 45

Overview 45
Technology & the European Market 49

Telephone History 50
Products and Services in the 1990s 62
U.S.-based Networks in Africa 67
Summary 69

Chapter 3 African Communities: A Case Study 71

Background of the Study 72
The Political Image of Africa 72
Respondent Traits 74
Study Limitations 77
Survey Results 78
Summary 89

Chapter 4 Africa and the New Information Supermarket 91

Overview 91
The African Mind 98
Strategies for Improving Services 102
The African Investor 102
Information Technology (IT) 103
Potential Benefits of IT to the African Economy 105
The Information Superhighway 106
National Dilemmas 109
Indigenous Habits and Foreign Corporations 110
Technology versus Performance 114
Education as a Communication Paradigm 117
Framework for Understanding the Technology's Financial Import 119
Networking in Academic Institutions 11
Information Sources 120
Installing Communication Infrastructure 127

Conclusion and Further Suggestions 128

Chapter 5 Exporting Communication Technology to Africa 131

Overview 131
Local Information Technology 133
Intelsat and its Services 137
Foreign Competition 139
Exporter's Procurement Policies 141
U.S. Telecommunication Export Policies
Exportation Problems & Prescriptions for Solutions 142
Telecommunication Services in Africa 147
Summary and Further Suggestions 151

Chapter 6 Telecommunication Policies in Africa 153

Overview 153
Regulatory Framework in Other Developing Nations 157
Privatizing the Industry 160
Telecommunication Status & Policy Update: Country Profiles 165
Further Discussions and Suggestions 179

Chapter 7 Using Information Technology for Grassroots Development 189

Overview 189
Information Acquisition Techniques 190
Biodiversity Conservation 192
Case of the Expatriate 194
The NGO Dilemma 196
Framework for Providing Telecommunication Services to Targeted Constituencies 197
Applying Relevant Communition Technology 201

Summary 206
Recommendations 207

References 217

Appendix I 223

Appendix II 227

Appendix III 229

Author Index 233

Subject Index 235

About the Author 243

List of Acronyms

ARPANET	Advanced research Project Agency. A set of interconnected computers that enable researchers to share ideas
AM	Amplitude Modulation (radio)
AAAS	American Association for the Advancement of Science. Headquarters in Washington, D.C.
BICIC	(*Fr.*) Banque Internationale pour le Commerce et l'industrie (du Cameroun)
BSA	British South Africa (colonial name).
CABECA	Capacity Building for Electronic Communication in Africa
CBS	Columbia Broadcasting System (USA)
CD-Rom	Compact Disk, Read Only Memory.
CERN	(*Fr.*) Conseil pour la Recherche Nucléaire
CFA	(*Fr.*) Communauté Financière Africaine. Currency used by Francophone African countries
CIDA	Canadian International Development Agency
DBS	Direct Broadcast Satellite
ECPA	Electronic Communications Privacy Act (of U.S.)
EDI	Economic Development Institute (World Bank)
FBI	Federal Bureau of Investigation (US)
FCC	Federal Communications Commission (US)
FTC	Federal Trade Commission (US)
GCE	General Certificate of Education. Awarded to secondary & baccalaureat students in some African countries upon successful completion of a written examination
HDTV	High Definition Television
IMF	International Monetary Fund

IP	Internet Protocol
INTELCAM	International Telecommunications Company, Cameroon
ITU	International Telecommunication Union
KPTC	Kenya Post & Telecommunication Corporation
MTD	Matchmaker Trade Delegation. A US-based company which coordinates trade relations between developing countries and companies in the US
MTS	Message Telecommunications Services
NAFTA	North American Free Trade Agreement
NAMIDEF	Namibia Internet Development Foundation
NGO	Non Governmental Organization
NTC	National Telecommunication Corporation (US)
NUCW	National Union of Communication Workers (Zambia)
OFCOM	Office of Communications (UK)
PANAFTEL	Pan African Telecommunications Network
PATU	Pan African Telecommunication Union
PPCC	Pre-paid (telephone) Calling Cards
PTC	Post & Telecommunications Corporation (Zambia)
PTT	Post and Telecommunication
RASCOM	Regional African Satellite Communications
SECAM	(*Fr.*) Séquence Couleur à Mémoire. An electronic color and scanning system used in Eastern Europe, Africa, and other parts of the world
SMATV	Satellite Master Antenna Television
STV	Satellite Television
TCP	Transport Control Protocol
UDEAC	(*Fr.*) Union Douanière et Économique de l'Afrique Centrale (Central African Customs and Economic Union)
UNDP	Unted Nations Development Program
UNZANET	University of Zambia (computer) Network
VSAT	Very Small Aperture Terminal
WALA	West African Library Association
WATS	Wide Area Telecommunications Services
ZPA	Zambia Privatization Agency

List of Tables, Charts, and Figures

Chapter 1

Figure 1.0 Communication as Activity 12
Figure 1.1 Language as Communication 13
Chart 1.0 Estimated Income of US Mass Media in Billion Dollars—1994 23
Table 1-0 VCR and Cable Use in Selected Developed and Developing Countries in early 1990s 27
Figure 1.2 U.S. Model for Industry-Driven Regulation 41

Chapter 2

Table 2.0 NTC, AT&T, Sprint Pricing. 4-5 Minute Residential Day Time Call versus NTC Saving 53
Table 2.1 Evolution of Major U.S. Long-Distance Carriers in Market Shares and Estimated Profit 54
Chart 2.0 1992 Estimated Profit & Market Shares among the Three Carriers 55

Table 2.2 Call-Placing on MCI World Phone Plan for Foreign Callers to U.S. 64

Chapter 3

Table 3.0 Frequency Layout on Communication Channels by International Respondents 80
Table 3.1 Equipment Importation by Business 80
Chart 3.0 Support for IT Transfer to Africa 84
Chart 3.1 Campaign for Creating and Managing Indigenous Telecommunication Industries 86
Chart 3.2 Product Ownership Versus view on Country's Participation 88

Chapter 4

Chart 4.0 Calls to the Host Country in 1994 122
Figure 4.0 UNZANET's International Connections 123

Chapter 6

Figure 6.0 Information Policymaking in Autocratic Regimes 155
Figure 6.1 Telecommunication Triad for International and Domestic Markets 161
Figure 6.2 Directions for Serving the Local Market 162
Table 6.0 Seminar for Communication Experts. Primary Education Campaign Strategies 181

Chapter 7

Figure 7.0 Internet Education. Connectivity Flow for Government-Funded Programs 208
Figure 7.1 Message Distribution Pattern for NGO-Sponsored Programs 209
Chart 7.0 The Internet; Proposed Country Bulletin Board 213

Introduction

The field of communication technology has been undergoing rapid changes in terms of equipment availability, use, and demand. Changes have occurred in the way communication technology research is conducted. Much research is now effect- instead of cause-oriented, catering to human capacity-building issues. By critically examining the uses of communication appliances including market analyses and political implications of their use, researchers have raised awareness on information-consumption patterns to a new level. This drift from scientific techniques to audience-oriented studies is not unprecedented, given a series of developments on the world stage—designed to bring international understanding and peaceful coexistence. The First and Second World Wars, the creation of international organizations to manage regional and global tensions, the reign of the Cold War and its dramatic conclusion marked by political conciliation between the Soviet Union and the U.S., the gap-bridging between market democracy and socialism symbolized by the reunification of East and West Germany, the struggle for civil obedience, especially in South Africa symbolized by the dramatic release of Nelson Mandela from prison and his eventual rise power, and the scramble for new markets in Third World countries, have all been documented by researchers and institutions.

The immensity of natural resources in the Third World justify the installation of new telecommunication networks for the maximization of group and individual business. Attractive tourist sites, a homely population, and an increasing number of airfields and cheap solar power all provide a conducive atmosphere for investment in telecommunications and other sectors. Computer networks and nodes are being used by agricultural, mining, and oil industries, to share geological and business information with experts in head offices. New markets in Africa may make that continent a viable business partner on the global stage. South Africa's information infrastructure which is as sophisticated as that of the First World, and its banking industry makes that country a world leader in providing customers with interactive banking facilities.

South Africa is one of the fastest-growing markets for cellular phone services and for the Internet. The size of the entire sub-Sahara market is estimated at 25 percent of that in South Africa, but it is growing at the rate of 75 percent to 100 percent per year because of a rapid influx of private companies into the region in recent years. But this growth is experienced only by major corporations. With an annual average revenue of $1,050 per telephone subscriber in the U.S., against $730 in Central America and $690 in Asia, and with about 125 residents per telephone in Africa, it is unlikely that many Africans will use a telephone in their lifetime. International lending institutions have been trying to reverse the status quo by sponsoring long term projects. The Bank, which sponsored the automation of the Zambian judiciary and UNCTAD which installed African trade points on an Internet server in Geneva, are among agencies determined to improve industries efficiency in Africa. In 1998, Chadian and World Bank officials held seminars and discussions regarding the establishment of a regulatory environment for the country's telecommunication sector. Certainly, this new order reflects a new level of cooperation between international donors and African telecommunication policymakers. More changes may develop from five situations:

a. More international providers will sponsor the installation of telecommunication systems, if African governments continue to demonstrate a willingness to implement policies that promote telecommunication operation;

b. More sophisticated networks will replace existing ones and major projects will use a combination of digital and analogue systems including satellite, cellular, and fiber optic technologies;
c. Telecommunication parastatal agencies and regional services operating beyond national boundaries will increase business opportunities for the private sector, with national and international shareholders competing to provide efficient services;
d. Workstation clientele will grow, if organizations continue using service providers and client-server application to transact business; and
e. Users will increase investment in the technology so will business opportunities and the demand for services. These conditions will increase opportunities for the exploitation of Africa's rich resources, thereby turning the once-upon-a-time "dark continent" into a world telecommuting center—an information world within an information world.

This new world order spearheaded by the G7, and promoted through U.S. Department of Defense's release of Internet services for global commercial use, the increasing availability of information products, and the demand by Third World governments for the establishment of technology-led institutional and educational capacity-building programs, have all brought new research challenges for communication scholars and policymakers. These entities must wrestle with the following imponderables: the effects of the communication order on monarchical and feudal systems of governing; the ability for poor countries and ghetto communities to participate in the new electronic information culture (a practice requiring a strong economy) without relinquishing their sovereignty and freedoms to economically viable groups and nations. Also critical is the ability for poor countries to acquire enough new technology in order to fully realize their development objectives that include economic independence.

Inasmuch as the so-called Information Superhighway or Infosupermarket can promote the global economy and peace through the sharing of business and cultural ideas, most Third World residents are financially and psychologically unequipped to operate in it. Eighty percent of Africans live below poverty. Eighty-five percent of those with reading and writing skills do not have access to modern information technology, so they cannot receive send, preserve or share information electronically. Although the IMF, the World Bank's Educational Development Institute, and its other filaments as well as other international donors have implemented programs where electronic information has been distributed from African nations capitals to the districts, although international agencies allocate additional funds for human-capacity building programs, most African governments have not been zealous when requesting funds to begin the installation of information infrastructures. Even when development projects are implemented using the funds, only the literate ones benefit first. These problems may create an elitist society and widen the knowledge gap between the Haves and the Have-nots—between the wealthy literate and the not-so-literate. Here, one is reminded of the colonial period when the new aristocracy—Euro-educated Africans and heathen-turned-Christians—nurtured better living standards than the rest.

But while the colonial administration (which produced that class of citizens) encouraged every African to receive a free education, the Information Revolution does not. Rather, to some, it is creating a new class of thinkers and consumers. Unless information products and services become available to all Africans, the continent's political and economic problems will be doubled in the next millennium. To avoid that neo-imperial condition, a realistic economic development plan free from *hidden agendas* and involving international and national policymaking and financial institutions should be implemented. International donors should collaborate with indigenous governments in designing long-term programs and in offering free information products and services. Otherwise, two social revolutions will occur that will worsen the prevailing economic and political problems already ravaging ravaging the lives of African residents: (1) a new virtual community of eccentric and highly informed people, and (2) an electronic colonization of the African mind. African governments struggling to adopt new principles of governance that include free expression need affordable communication systems and services to assist them with the new world order. Equipment and services should be exported primarily to assist these countries in realizing their economic goals, and in fostering international peace.

Although the issues discussed in this book may appeal to a *public relations* or *corporate communication audience,* it tries to establish a broad realistic description of the commercial communications function.

The book does not have stone-carved answers to problems of exporting communication technology. Nor does it completely describe market conditions and telecommunication policies. With the constant creation and widespread use of information technology, no printed document can provide accurate information on the production, marketing, and consumption of the goods. Hence, it is more appropriate to view this book as a tutorial package for companies, governments, new-comers into the world of communication technologies and their role in assisting African countries toward the realization of their development objectives. Also, it provides a general analysis of the implications of exporting products to developing areas. Readers should not consider it a handbook of current facts about telecommunications in Africa since regulations and statistics constantly change. Additionally, given the arbitrary political climate there, the exportation of information technology, procurement policies, and market cannot be stable.

Exporting technology should not be construed as removing something from one place to another place. In this book, it means using specific information equipment developed in one place to promote development in another place. Certain cultural and philosophical habits abound in the process, and indigenous policy makers and trained experts are expected to direct the application of the technology. This process could prevent military coups and unstable governments, by increasing job opportunities for idle young men who habitually join the military for employment security reasons, to control civilians, or receive bribes, instead of monitoring peace and justice and protecting the nation. Imported technology should help curb corruption and stabilize Africa's economy during the Twenty-First century.

Communication technology has four other components: research/process, market, profession, and practice.

Research. Communication technology is a system, which can be tested and retested, that is, studied across time and space. Researchers are charged with the responsibility of gathering information, designing programs, implementing action, and evaluating results. Hence, communication technology can be seen as having a management function.

Market or ***public*** refers to any identifiable group or individual who the technology targets. It is an institution (public or private), employee, employer, media, place, or consumer.

Profession refers to Communication instructors scholars, and practitioners trained individuals who help other groups and individuals acquire, use, maintain, and share communication equipment. Professionals usually operate in or for organizations. With the unprecedented demand for more new information technology and the growing interest in its financially rewards, more communication experts have been setting up independent consulting firms.

The ***practice*** of communication technology is directed by the market it responds to and the various institutions it serves. The changing characteristics of the market—different cultural habits, new generations and their tastes, increasing popularity and availability of information equipment and its easy operative techniques—have made the practice of communication technology ever more challenging.

There are seven chapters in this book divided into two parts and several non-sequential sections. The first part, which defines and describes some of the modern technologies (e.g., Internet, HDTV, facsimile), targets college-level readers. Undergraduate degree programs handling communication studies, international, and economic affairs in academe, and the private sector business and customers) should find valuable information here. The second part examines the role which information technology plays in economic development and cultural pluralism; it describes the political, philosophical, and cultural role played by modern technologies in economic and cultural imperialism and provides a market analysis of the prospects of transferring or exporting and utilizing communication technology in Third World countries.

Chapter 1 defines communication technology with examples, presents a communication philosophy, and analyzes the socioeconomic potentials of new technologies: the satellite system, beeper, teletext, computer (Internet), and high definition television.

Chapter 2 presents the history and activities of U.S. long-distance telephone companies and points out their potential for attracting the African market. It highlights AT&T's Africa One program in Africa and ways in which the program could become beneficial to both the company and African people. Also highlighted are activities of regional African telecommunication companies like RASCOM, INTELSAT, SADCC, PANAFTEL, and Afro Network, Inc., a Delaware-based corporation poised to assist African countries in modernizing their services.

In Chapter 3, a case study, the perspectives of African residents and embassy staff on the transfer to and use of communication technology in the continent are analyzed.

Chapter 4 lays the groundwork for executives who intend to do large-scale business with or for Africans by analyzing how Africans think. It also provides an international perspective on the information superhighway and describes its effects on the social, political economic, and educational predicament of Africa. It raises a number of concerns about Africa's participation in the new world communication order and proposes significant measures which African policy makers and international industries must adopt in order to ensure Africa's continuous membership. Traditional /indigenous equipment (the "talking drum", flutes, and bells) and their ability to transmit specific messages in ancient and modern rural communities in Africa are examined.

Chapter 5 looks at U.S. telecommunication export policies and their potential effects on U.S. trade relations with Africa. It shows how the exportation of too much telecommunication equipment to developing countries can promote neo-imperialism, and how limited importation could encourage the growth of private industries within developing countries and eventually expand competition in the use and manufacturing of such products. Financing and equipment sustainability are among other issues discussed. Regarding issues related to telecommunication services in Africa, the author shows how outdated equipment, poorly trained personnel, insufficient finds, and a lack of interest in improving government controlled telecommunications could permanently erode the African market, and make dialogue between international companies and African government officials more difficult.

Chapter 6 describes telecommunication policies designed by African policymakers for their citizens and for the international organizations they do business with and provides a framework to be used by new foreign companies and progressive institutions in adopting synergistic policies. It further proposes policies, which can help both parties in maximizing benefits.

Chapter 7 examines the relationship between traditional micro-economics theory and development, how an African information industry undertaking the production distribution and preservation of information for use by trained entities on the continent can be cost effective, generate significant change in the social economic arenas, and improve the living standards of communities that use information technology.

Writing the book was a new challenge for me. My trips to international libraries, bookstores, newspaper kiosks, government offices, and private corporations, and analyses of media broadcasts have increased my appetite for the field. Some of the information I found in computers and news magazines was often filled with editorials, prognostics, and biased statements about informatics, telecommunications and their markets in global society. Although computers had more current information than the periodicals and books, much data drawn therefrom did not serve the immediate expectations of the scholarly text I had set out to produce. However, the field of *communication* or *information technology* is market-oriented, catering directly to public needs. And that is what this book may have accomplished.

Exporting Communication Technology provides a general analysis of the political, economic, economic, cultural, and educational effects of using foreign technology toward human capacity-building in developing countries in particular, Africa. This book considers communication technology as playing four main roles: entertainer, generator of infrastructure development, regulator of business transactions, and persuader of human thought. It presents civil society as a state of mind and as a place capable of progress and describes market conditions in Africa and the U.S. It also examines the role to be played by mainstream providers, expatriates, retailers, and Third World governments in fostering international business.

Acknowledgments

I am grateful to Professors Berishetta Merritt, Robert Nwankwo, and Sulayman Nyang of Howard University's School of Communication and African Studies Department, Dr. Christopher Simpson and Professor Hamid Mowlana of the Journalism Department and School of International Studies respectively at American University, and Dr. Andrew Moemeka, Chair of the Communication Department at Central Connecticut State University, for their suggestions and advice toward the completion of this book. I appreciate the spiritual support from my wife Manigha, my friends, especially, Tim Carrington and Mark Woodward at The World Bank office in Washington, D.C., Ted N'Kodo and Valentine M'Barga of the African Development Bank, and my entire family during the preparation of this book.

Emmanuel K. Ngwainmbi
Washington, D.C.
January, 1999

Part I

Instructional/Academic Overview of Infotech

Chapter One

Communication Technology

Can you imagine any human community today without modern telecommunication technology? Certainly, Third World people can. Elsewhere, people use and abuse telephones, beepers, fax machines, and the like, because they are cheap, available, and user friendly. Over the decades, communication in the developed world evolved into an institution. In the international community, specifically Europe and the United States, the field has been enhanced by innovations in technology: the telegraph, print, the telegraph, print, telephone, satellite dishes, high-definition television, fax machines and fiber optics. Fiber optics in industrialized societies increased and sped up communication. Today, people can interact anywhere in the world through the Internet, as long as their computers are connected to the PC servers. Companies, governments, public and private institutions use technology for entertainment, profit-making, fast and easy interaction, educational advancement, and cultural exchange.

To what extent can communication be observed as a technology, an activity, a concept, or all three? Different schools of thought have advanced theories relevant to the term. One would not be too generous in considering communication an activity, a concept or a technology. As an activity, communication makes things happen. An action occurs, thereby provoking a reaction. A message sent is received, and the recipient (person or machine) is stimulated by the message to respond. This response may be noise, movement, listening, silence, or indifference. (See Figure 1.0).

Communication begins in the mind of a person or other living thing before an action occurs. An idea is conceptualized and put into action. We manifest the senses of feel, taste, touch, smell, and hear. Animals have and manifest at least four of the five human senses; but unlike humans, they communicate their emotions instinctively, and most often receive an immediate reaction. When a dog bites a man, the man screams immediately. When a dog hurts its own leg, it winces automatically. For humans, however, communication and action are more complex. Feelings differ in the way people perceive an event, an object, an idea, or a practice. Man's perception of man's culture differs from how other people perceive it. The words and meanings we communicate in our language differ from the language and meaning in other cultures. Ferdinand de Saussure the Swiss linguist, Edward Sapir, Noam Chomsky and other linguists have successfully argued that language is a part of culture. Language enhances communication. A feeling stimulates an idea and the individual provides a language for that idea. Through language, an interaction occurs (see Figure 1.1). This process produces communication.

Figure 1.0
Communication as Activity

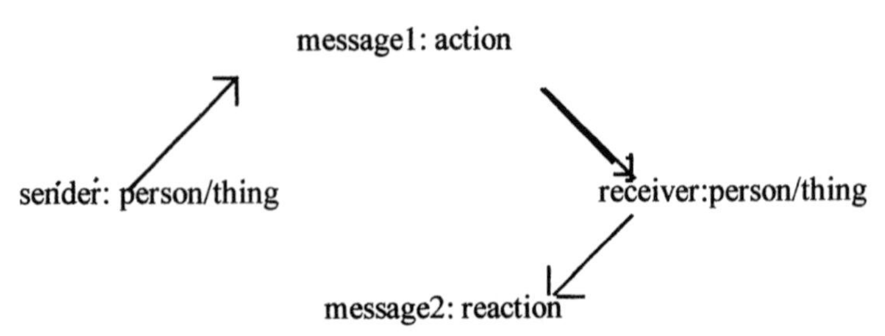

Message 1 refers to the action from the sender or source. Message 2 is the receiver's reaction to Message 1. *The response can be immediate or delayed.*

Figure 1.1
Language as Communication

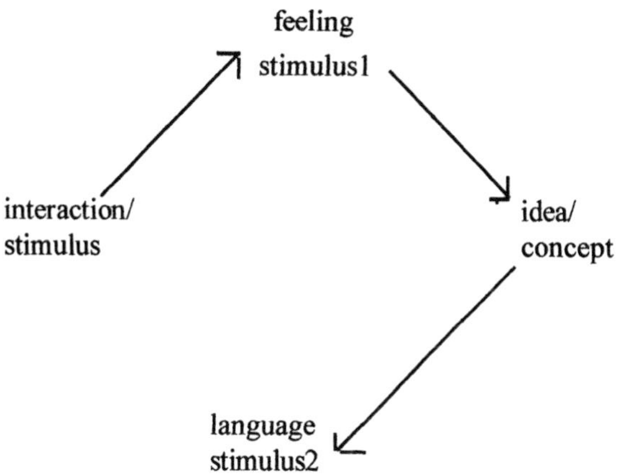

feeling
stimulus1

interaction/
stimulus

idea/
concept

language
stimulus2

Telecommunication versus Communication Technology

Every student of communication understands the conceptual definition of telecommunication. However, concepts of its functions differ considerably depending on one's geoethnic location and economic standard, and exposure. The "primitive" individual has a limited exposure to telecommunication systems, hence a limited knowledge of their use. This is unlike the individual in an industrialized environment, who in addition to using communication systems for entertainment, applies them toward creating economic opportunities and maximizing profit. Typically, for the "primitive" entity, communication forms mediate social messages, send news, spread information at public gatherings, and produce music (using drums, tambourines, and flutes). Such systems, albeit familiar and readily accessible by the rural entity, is unexportable because they have relatively limited market value compared to electronically manufactured systems.This book analyzes the nature and effects of transporting telecommunication equipment and services to economically underprivileged areas.

Since we understand that "tele" means "at a distance", and communication is the sending, storing, receiving, and conceptualizing of messages, we cannot dissociate the function of telecommunication from society's fabric or from its values. Debates on the socioeconomic functions of telecommunication have created several contexts for analyzing it. According to a February 9, 1988 Commission of European Communities report, telecommunication has become a conglomerate global sector comprising the management and transportation of information. The Commission oversees the activities of sub-institutions: companies, industries, governments, and groups. Based on the other context, telecommunication functions effectively where people have the freedom to make choices in, an open (free) market economy, a civil society.

Civil society is synonymous with "political society" and political freedom. Freedom suggests individual rights. Political society is a place where human conduct is determined by defined accepted principles, while political freedom implies having the freedom to choose a course of action. Seligman (1992) has noted problematic relationships between the private and the public, the individual and the social, public ethics and individual interests individual passion and public concerns (p. 5) in defining civil society. Philosophical differences prevail between civil society and political society. While sanity, public interest and coercion characterize civil society, political society is fragmented—the driving forces are: self-assertion, self-determination, self-empowerment, individualism, and independence. Hence, agents of individualism like technology and money can and do shape a political order and they are motivated by profit-oriented incentives.

In a century charged with technological imperatives, rapid changes and global communication, telecommunication itself does not discriminate between political society, political freedom, and other social indicators; rather, it complements them. Although technology and communication infrastructure have accelerated the destruction of the ecosystem information technology, as conceived, has always been intended to contribute to the improvement of all levels of human life. Through technological innovations, the quality of human life is highly facilitated by both wireless and wired telecommunication systems. Communication has been faster since the telegraph, telephone, television, VCR, electronic cable, computer, beeper, and alarm systems were invented, making doing business, for example, faster and easier for everyone using them.

Communication as Technology

As the world approaches the twenty-first century, human institutions are creating communication models and building infrastructures on a parallel course with the ever-increasing demands for better living standards. People are choosing new communication technologies because they are accessible, reliable, portable, increasingly private and have easy message-storage/message-retrieval capacity, and are easy to operate. But what is communication infrastructure?

The Institute of Culture and Communication, East-West Center–Honolulu, Hawaii has described communication infrastructure as an agent of development. Modern buildings, telephone wires, satellite plants, fiber optics, television and radio stations came into being as people built houses and lived together. Skills and training are required for the operation of telecommunication technology and infrastructure. Be convinced that an operator of a radio equipment and/or radio station must have undertaken rigorous training. This means learning the language and skill of its creator. Learning or acquiring knowledge in a foreign culture is economically advantageous; a trained individual acquires skills necessary to market him/herself anywhere in this rapidly uniting world. A fax machine operator is more likely to get a job, or communicate faster with anyone in the world, than an academician who has not been exposed to any telecommunications technology.

But what is technology and how is it related to telecommunication? In some universities across the United States, the Telecommunication Department is different from the Mass Communication Department in terms of course titles and degrees offered. In fact, telecommunication is a branch of mass communication in that large audiences receive messages from distant communication channels/sources. The idea here is, mass communication is commonly practiced in homogeneous societies. People in African villages and townships spread and share information using drums, gunshots, speeches, and public gatherings. These are interpersonal modes of communicating which offer valued skepticism, not on the function of communication but on its effectiveness.

Can the act of communicating generate a common reasoning pattern? Is truth perceivable only through a complete process of communicating from encoder to decoder and vice versa through a unidimensional process? Can members of a particular class use common channels and principles to produce their own reality?

The answers to such questions lie in our articulation of two premises:

1 The morally positive characteristics of parties who seek a rational advantage through communication systems, and
2 the situational constraints under which those parties choose policies for cooperating with other parties.

Jurgen Habermas, German philosopher of the Johann Wolfgang Goethe-Universitat, argues that because citizens are moral persons with a sense of justice, they can form their own conception or cultivate dispositions in a rational manner (Habermas, 1995, p. 112). Here, citizens of any society are seen as a part of modern technology by virtue of their ability to make rational decisions about their lives and environment. Hence, designing modern electronic equipment can be associated with satisfying human needs.

As with other civilization promoting agents, communication technology seeks to convert the world into a global civil society where information is available to everyone. This state of existence has been conceptualized by Habermas, Heidegger, Kant, and other European social scientists. The philosophers posit that an individual is created to do what is equally right for all persons. Rightness presupposes effective communication between the encoder and decoder. Whether it was expressed overtly or covertly, the interaction must be completed and understood by the recipient. Mowlana and Wilson (1990) have defined communication as a development agent—the "unfolding of knowledge, a transformation from being to becoming". Through transformation society creates signs and actions that translate into action. The road from "being" to "becoming" or from "signs" to "actions" is paved with equipment and service. No communication can successfully occur without a means or tool, an effect or service rendered. Therefore, communication does not take place for its own sake, but to satisfy a need. Information exchanged to cause an action must be carefully designed and preserved.

As developed countries continue to assist Africa in preserving information on health, biodiversity sustenance, and cultural activities through video-cameras, audio and video equipment, and print material, it is imperative that the entities they are assisting comprehend the operational meanings of the material being conserved. More importantly, recipients should formulate their own operational meanings of telecommunication because at present information on crucial matters flows only in one direction—from developed nations to Africa. Also, due to a lack of fast, modern, and efficient telecommunication infrastructures in Africa, interaction between African institutions, scholars, and researchers on cultural and development issues is highly disjointed. Africans continue to rely almost exclusively on information from abroad to attain their development objectives. This has perpetuated a pattern where developed nations are in a position to make decisions about crucial issues that affect Africa's present and future: famine, health, politics, and economy, based on information cabled through satellite systems. Thus, information technology is transferred to developing countries without enough understanding of its users' needs or a clear articulation of the support needs of the information systems. This mentality is imperialist, and has heavy negative consequences on developing countries. Lee (1986) observed that the introduction of advanced communication technologies such as computer networks and satellite broadcasting systems to many Third World countries has worsened these countries' cultural and financial dependency upon the advanced nations supplying them. The main problem here is, no superstructure—privileged class—can willingly share power-sharing with targeted groups is essential to progressive communication.

The international community appears to be oblivious to socioeconomic and political transformations in Africa when transporting telecommunication services and equipment. Similarly, information about Africa in the broadcast media in industrialized countries is fragmented—lacking in substance. Frequently, development and conservation practitioners operate projects in Africa by trial and error, simply because of limited communication between them and target groups. Donor institutions do not seem to understand the groups they work with because they do not operate through people-oriented mechanisms and people-managed machinerymhuman infrastructure, culture, society and telecommunication equipment. Power sharing among target groups is certainly essential for an effective conservation and improvement of Africa's biodiversity. Decisions about how to implement mass communication systems must be shared, a collaboration between the suppliers and the eventual users. The development and conservation community must be willing to invest time and financial resources to improve targeted areas in the continent. Only then can effective communication be said to have taken place between donor and beneficiary. To contextualize this, a brief tutorial on forms of communication technology follows.

Computers as Communication Technology

A computer can be defined simply as an electronic system which calculates, stores, and prepares data for use. Computer use in developed countries is an activity that is associated with success. The material development of countries relies basically on the operation of computers.

Businesses that use computers harvest more profit than those without computers. Computers make possible immediate communication between sender and receiver, and transactions may take five seconds or less to complete. Large volumes of data are transmissible through circuits built into a computer network. Information can be retrieved from almost anywhere in the world, providing the source and receiver have access to the same network. A network comprises the electronic connection of two or more computers for the transfer or exchange of data. Many computer networks are in use all over the world, but the most significant of them in terms of social interaction and international communication is the Internet, a collection of networks connected by routes—machines which make it possible for a computer on network A to communicate with computer B or others on the networks. The Internet, which as of 1996 has connected approximately about 3 million computers all over the world, is favored by academic and business institutions because its language is easy to understand and is accessible to almost everyone in industrialized societies. The IP, or Internet Protocol, communicates between two computers in different places irrespective of their hardware components. Network, data link, transport and application layers of the IP facilitate the communication of information or messages from computer to another. However, "a standard language or protocol is required to specify a set of rules that would make communication possible" especially when interconnecting networks use different architecture (Orondo, 1994, p. 39). Each computer in the Internet has a specific identification—a name with dots—which facilitates the communication of messages at any time to anyone connected to the Internet.

Certainly, the computer is an effective advanced technology and it has upgraded the quality of life where it is used constantly. Computer literacy has become a new world order as major corporations seek to ensure that all potential employees receive computer training to meet the demand of increasing technological innovations. In America, computer lessons are given to youngsters at all levels, from elementary to university, and American academic institutions have devised innovation programs to foster computer training. Even toys have computerized systems which influence the thinking procedure and activities of children from very young. The successes of this technology is enjoyed mostly by developed countries, mainly because they can afford it.

The role of computers in developing countries was critically examined in the 1970s by the United Nations. A 1971 UN report: "The Application of Computer Technology for Development" predicted:

> Computers will play an increasingly important role in developing countries which intend to participate in the world economy in ways other than the supply of raw materials. Developing countries will find computers a necessary ticket of admission. The next decade should see developing countries even more active in closing the computer gap (UN, 1971).

That prediction was made over two and a half decades ago. Yet, since then, the wholesale export of computer technology to developing countries has proven to make only marginal contributions to their economic growth and development. Although the language of the software is primarily English (used in most parts of Africa), its instructions are sometimes too technical for those operating the system. The databases do not contain local documents or signs commonly understood by most Third World people who have a minimal education. Other obstacles could prevent this group from fully exploiting the computer database, or obtaining first class information. The difficulties are:

1. Western technology cannot be absorbed wholesale.
2. The computer software systems could not be developed in most African countries because of a lack of experienced or trained personnel. Although some Nigerian institutions are experimenting with the manufacture of microcomputers, their less sophisticated equipment and the high cost of raw materials do not guarantee the manufacturing of durable, quality computers.

3. Being a tropical continent with excessive humidity, dryness, and dust, even well-built computers will easily be damaged and will need repair, whereas relatively fewer repair technicians exist. Technology is required to insulate computer hardware from the elements of the tropical climate.
4. Systems are expensive, hence hard to implement and maintain in that impoverished continent.
5. There is limited electricity. Even the electric companies managed by private owners do not function properly. Frequent power outages easily damage computers.
6. Few libraries, schools, and bookstores are equipped with computers, and the high cost of foreign publications—printed and video materials—makes it difficult for even those institutions which do have the hardware to import them. Even institutions that can afford to import computers, like the University of Zimbabwe and schools in South Africa, do not yet have direct access to current information about new publications. This information shortage can be blamed on both the government and public sectors, which place other priorities above information acquisition. The incentive for aggressively seeking new information has been stifled by African leaders, indeed dictators, who employ undercover police officers to arrest people suspected of seeking or spreading certain types of information that the dictatorships find threatening. Even trained news-gatherers—journalists and media practitioners—are often arrested, detained and convicted for gathering and spreading information not processed by the government. This makes the information institution vulnerable and less than desirable to the average student who cannot gauge. Also, the concept that public service messages in the rural areas can only be disseminated (by special individuals from the King's court) (Ngwainmbi, 1991, 1995a) makes information a privileged commodity. Thus, the application of and accessibility to electronically processed information could be difficult and scarce for most African residents in the twenty-first century.

The twentieth century has, however, witnessed huge manufacturing and use of telecommunication systems, from the telegraph and telephone, typewriter, computer, VCR, satellite dishes, radio, television to video text, usenet and teletext. With the advent of radio, telegraph or telephone, non-pictorial messages can be transmitted between distant places within seconds. The receiver can conceptualize and respond to the message he/she hears by wire, without seeing the sender. The creation of the video text and tele text, has facilitated searches, information-creation and information-sharing, cyberspace, and high definition television (HDTV).

Satellites

A satellite is an electronic system launched in space carrying transponders that identify and transmit signals sent from the earth. By the year 2000, a new satellite will be able to carry 40 or more transponders. Each transponder will carry 1,000 telephone messages at a time. A technological innovation in the last few years, the transponder has replaced the less effective microwave transmission whose signals are manipulated around mountains for audio and visual messages to be transmitted. The transponder is superior to microwave transmission in that it provides a direct signal into space and back to earth with no ground interference. According to Wilson (1995, p. 250), U.S. communication satellites cover about 22,300 miles around the equator and rotate at the same speed as the earth, as opposed to non-U.S. satellites. With the ability to send signals to thousands of miles, the satellite system can provide long distance communication services at a low-cost, especially to African nations, which are relatively poorer than other countries. Some owners of satellite dishes and cable TV sets have been using this technology for many years in watching live and taped programs worldwide. For example, West African viewers witnessed Mandela's inaugural ceremonies as well as the 1990 and 1994 World Championship Soccer matches from Italy and U.S. respectively, President Clinton's visit to Africa, and presidential election campaigns in America, via satellite.

High Definition Television (HDTV)

In a description of HDTV, Wilson (1995) has reported that the U.S. uses a system of picture elements, arranged in horizontal lines consisting of 525 lines and that the TV screen exhibits 30 of the 525-line still pictures, which give the illusion of motion (p. 252). The Japanese invented the High Definition Television (HDTV) system that broadcasts 1,125-

line television pictures, and converts the analog signals originally used to broadcast video into digital signals (p. 252). Cable companies in the U.S. can deliver more channels to consumers with this technology than before. However, HDTV is still under experiment. According to some experts, it will take at least 5 years from 1995 to fully develop the technology. This technology, which transmits many digital signals and allows a viewer to receive or view many channels on one screen at the same time, may for decades to come only be used by the rich in Africa, because it will be very costly. The cost of converting from regular TV to HDTV viewing is very high.

Charged with subliminal messages, these technologies promise a new definition of culture and a new way of thinking. From an economic perspective, HDTV will yield huge profits for advertisers and TV companies, as more people will be charged for the additional channels they watch. Since a 30-second ad on a national prime-time U.S.-based television network costs approximately $100,000, even higher charges will be levied on advertisers using HDTV. It would be interesting to know who would charge whom for HDTV programs in Third World countries, and whether Third World advertisers would be able to sustain the purchase of international air time for commercial purposes, given that the majority of Third World audiences live below the poverty level. However, through HDTV, most viewers will appreciate other cultures.

While HDTV may become a less lucrative commodity for the Third World market, computers and long distance telephone businesses may bring more profit, due to new sales incentives and the fierce competition between long distance telephone companies, (especially AT&T, MCI, Sprint, Excel, and NTC) with access to markets worldwide. To gain a better understanding of why telephone companies stand a better chance of exporting and selling their products to Third World countries, read the next. Chapter Two identifies the products and analyzes the companies' market strategies targeted at U.S. and international consumers, their rates and their products.

Television

The origin of television has been traced to Russia. In 1923, Vladimir Zworykin developed a camera tube. Britain was, however, the first country to begin regular TV broadcasting in 1936. In 1939, Americans were first exposed to TV at a World Fair in New York. By 1948, CBS and NBC

nology was further advanced in 1951. The microwave provided coast-to-coast TV programming for audiences for many years before transponders were invented.

As a communication medium, television has expanded its coverage of social and political events beyond American borders, "exporting" huge varieties of American culture to other countries. A 1994 report from the U.S. Government Printing Office shows television as the second of seven media which yield an annual income of approximately 25.6 billion (see Chart 1.0). This clearly shows that TV is second to the 5,000 daily newspapers in the U.S. which yield about 44.3 billion annually due to their aggressive direct reporting approach,their distributing, and marketing techniques. However, TV is the most popular information and entertainment channel. Even nationally syndicated and non-syndicated

Chart 1.0
Estimated Income of U.S. Mass Media in Billion Dollars - 1994

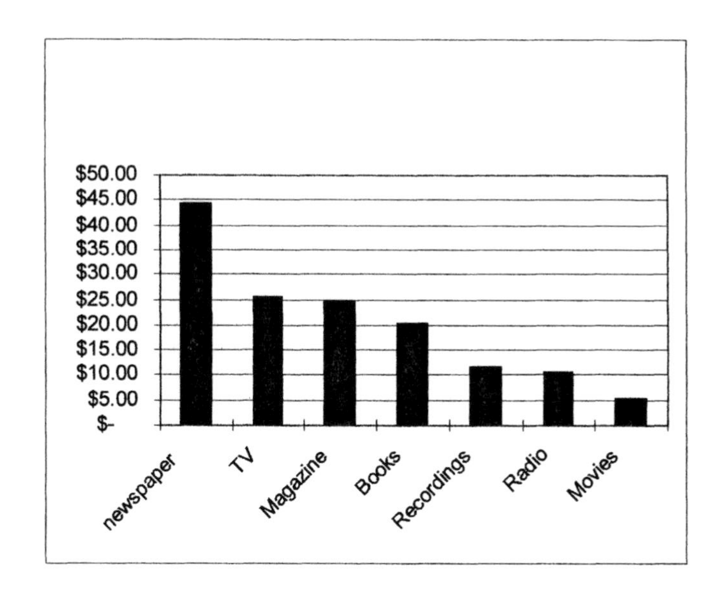

Source: *U.S. Industrial Outlook*, 1994. Washington, D.C.: U.S. Government Printing Office

entertainment channel. Even nationally syndicated and non-syndicated radio programs elevate radio to only the sixth most available commercial communication medium. The creation of TV services and VCRs (video cassette recorders) in America, Germany, and Japan has revolutionized public viewing habits. Although fewer Americans now watch regular prime time television, its viewership will likely increase with the invention and availability of high definition television, since viewers tend to patronize new technology in order to satisfy their curiosity. The things that increase entertainment value arouse curiosity and raise purchasing interest. In Peter Benjamin Seel's 1995 study of HDTV in Japan, the U.S. and Europe, Seel asserts that HDTV users are providers because they decide whether a commodity should be marketed and the extent to which it can be marketed. Although government regulates telecommunication operation, the industry defines the success of telecommunication products based on market research.

Cable and VCRs

This section highlights some of the technological, cultural, and socio-economic issues which producers and consumers of cable and VCR appliances sometimes overlook and which can slow services rendered to the users. It also describes how cable and VCR function, in order to inform individuals interested in purchasing or using them.

Cable network programs are transmitted over satellite to the system headend (Gross, 1997, p. 69) and cable receives video information through wires, rather than over the air. The video cassette recorder (VCR) is an electronic device which can be used to record over-the-air programs on network TV, cable, or from a radio cassette. While cable receives information, the VCR records it. Viewers can be instantly exposed to live programming. Subscribers to cable services can watch CNN and other major news networks worldwide. The Pope's visits, the fall of the Berlin Wall, political changes in Russia, the murder and sex scandals emanating from American communities, and images of Africa, are all cabled to millions of viewers around the world. Certainly, this technology will enlighten people in any culture.

Technological Aspects of Cable and VCR

Cable signals are delivered through the air from broadcast antennas and from satellites which place signals on wires. These wires are either buried underground or connected via telephone poles. Cable TV has a capacity of being a two-way communicator, with a variety of channels. The average American cable TV set has about 200 channels. A cable system needs a head-end that receives all signals. This site is then sent to subscribers. All signals come into the head end and undergo processing in order to be placed on a particular channel for the customer. The material gathered from the head end is then placed on a coaxial cable and the head end sends signals through the air to hub microwave dishes which, in turn, send signals to homes that have been wired. Cable TV programming consists of pay-per-view, basic cable, local programming and auxiliary or international services.

VCR technology has the following features: recording, fast forwarding, rewinding and stop, all maneuvered with the aid of a button. Some VCR clocks can be used to record the time the program is taped. Some VCRs also have a built-in timer which enables a user to record one program while another is being viewed. The VCR can record over-the-air programs. It also consists of the following features: still frame, which can enable a user to hold one frame on a screen for examination. The frame-by-frame advance facilitates the viewing of one tape at a time. Slow motion requires playing back at a slower rate. Cable equipment is more complex and more expensive than that of a VCR. While cable involves huge, heavy and expensive structures, the regular VCR is less expensive and portable. Prices of VCRs vary, but the overall cost of the cable dish is higher.

Socioeconomic Differences

Like cable television, the VCR has made movie-goers concerned in that they pay more money to watch movies in theaters than to watch the same movies broadcast later on cable. Movie-goers are sometimes confused about whether to wait and watch a movie on cable or to spend more money going to the theater. Whatever the perception, it is more expensive to have cable TV than to go to movies. Also, cable advertisers struggling to reach a large audience are unsuccessful because many Americans switch from cable to VCR, especially during winter. During summer, cable rates drop dramatically to attract a waning clientele that is more inclined to being outdoors. Cable TV which has pay-per-view programs faces difficulties because many people subscribe to cable in order to watch special interest programs. At the end of the program, some subscribers request discontinuance of cable service. Unlike cable, which relies on commercials to remain in operation, subscription TV relies on subscription fees. Pay-per-view programming has been picketed by entities like SMATV, DBS and Loner Power Station for profit-making; yet such programming has persisted and has even increased its market. Cable subscribers can obtain pay-per-view (PPV) programming two or more times a month, depending on their interest in watching the program. The dominant program is sports (boxing and football). Sports fans have subscribed to heavy weight boxing matches even when such matches cost $5O per view and only last an average of one hour. Conversely, other technologies like satellite television (STV), videocassette, video discs, and CD-ROM, which sell the same shows that cable airs, are in less demand because cable programmers advertise their shows more frequently than the latter. Its long-time relationship with other entertainment industries gives STV an edge over VCR which only becomes efficient when used with a TV set. Besides, its customers are accustomed to paying for the services it renders. VCR use has led to a decline in both cable and normal TV viewer ship. As more people purchase VCRs, regular TV advertisers may be losing a lot of money because a VCR owner can record a concert and delete the commercials in the concert program. A VCR user can also purchase whatever video tapes he wants. Thus having a VCR can be better for a consumer in a slow economy.

VCR use has increased beyond U.S. borders. Dominick, Sherman and Copeland (1993, p. 173) report that 70 percent of households owning TV equipment in Australia had VCRs; the rate was 66 percent in Canada and Netherlands and 40 percent in Mexico. Even in the Middle East where foreign religions, social freedom, and mass media activities are heavily censored, VCR use has increased dramatically among families with TV sets (see Table 1.0). VCR is fast becoming a major telecommunication product worldwide. Data collected in 1991 show a higher percentage of homes in Saudi Arabia owning VCR than in Switzerland, a democratic country. The total absence of cable household users in that country clearly suggests government censorship of information and media use. Any cable use may infiltrate the minds of the Muslims and threaten Islamic universalism.

Despite the conspicuous absence of African countries in this study, it

Table 1.0
VCR and Cable Use in Selected Developed and Developing Countries in the Early 1990s

COUNTRY	% of Households with VCR	% of Households with Cable
Australia	72	0
Brazil	45	0
Canada	66	82
Germany	30	33
Great Britain	60	7
Japan	68	18
Saudi Arabia	60	0
Poland	5	0
Switzerland	58	75
United States	72	60

Source: Dominick, J., Sherman, L.B., and Copeland, A. G. (1993) *Broadcasting and Beyond: An Introduction to Modern Electronic Media* (New York: MCGraw Hill, p. 173)

must be noted that about 80 percent of Africans abroad, especially between 18-45 years old, own VCRs and 75 percent of them ship VCRs to Africa per year. This poses several serious problems. Because most of the video cassettes being shipped contain U.S.-produced movies, African countries struggling to maintain their culture must produce and promote their own videos to compete with the massive influx of foreign videos. Viewer support for "home brewed" programs may drop as viewers may show a preference for more time watching foreign videos. Secondly, African video crews do not have the same amount of funds like European and American partners to prepare complete, quality videos, and yet they are obliged to compete with the foreign video industry.

Cultural Aspects

Satellite dishes placed anywhere can provide cultural exposure to many people in the world. With the aid of cable or a satellite dish, Americans can and do learn about events outside America. Live political protests in South Africa have been captured by satellite and transmitted to American homes. People in remote areas remain ignorant of international affairs where there are not satellite or cable facilities. People learn about different people in different areas, and watch different events with the help of pre-taped video cassettes and VCR sets. However, information on VCR can only be played back and that makes the information suspicious because the recorder can change some things to suit his interest.

The Internet

The Internet, or the net, is a rapidly growing information system that was started in the mid-1970s by the U.S. Department of Defense. Before the Internet became a public commodity, it was used for military activities. U.S. Department of defense had sought to build an advanced disaster proof system which would also supersede Iraq's system during the Gulf War in the early 1990s. The TCP/IP had routers to facilitate its chain of command. The U.S. tried to destroy parts of the network—cut off Iraq's vital information links, but failed because TCP/IP could bypass the nodes that had been put into action to reclaim the network, hence protecting itself.

The Gulf War was a chance to experiment with other aspects of the ARPANET—a set of computers around the world. Because of its ability to report things in real time, the Internet became the military's ally. This system allowed specialists and other interested parties to discuss events of the war as they were taking place. Before it hit the civilian world, the U.S. Defense Department linked computers from around the world. ARPANET was soon introduced to universities and corporations as a source of public information. Spreading like a virus, the Internet had approximately 15 million users in nearly 70 countries by 1993. The Conseil pour la Recherche Nucleaire (CERN), an international organization whose members span across the world, wanted a fast means of mediating messages. In 1989, Tim Berner-Lee led a team of scientists in developing a means of sharing information on the Internet. This led to the founding of the World Wide Web. In 1993, the National Center for Supercomputing Applications (CSA) introduced Mosaic, a software program that displayed Web graphics, text styles, hypertext and other files on the same page. Several versions of Mosaic were later released as noncommercial software. The versions included Windows, Macintosh, UNIX, Sun, and Silicon Graphics work stations. Web users were thus able to add their own information on the web and by 1993, 100 official web sites were available. After Mosaic was released, more people started using the web.

The Internet has broken the geographic and transportation boundaries set by the facsimile, TV, radio, beeper and other telecommunication systems. It can be seen as a messenger of evil or good will, depending on the user's objective. In industrialized countries, the technology has brought new challenges for law enforcement institutions and new opportunities for crime. It has become a community, a place where people are predisposed to committing crimes. For example, a public bulletin board called alt.sex.bondage was the place where a murderer seduced a female Internet user to a face-to-face meeting, then killed her. Electronic relationships provide an ability to reinvent oneself, to explore one's imagination. The dialogue which goes on between strangers introduced through each other's e-mail address is spontaneous and creative. Although on-line users become obsessed with its dominant-submissive role-playing that promises happiness but never provides satisfaction, the Internet has generated a new level of social and economic consciousness—it has created a new ethical standard. It distracts the mind from contemplating evil by occupying peoples' time. Researchers and students no longer need to undergo the ordeal of locating material from card catalogs, or purchasing news magazines and other print material. A simple subject, title, or word search provides updated information.

Although the Internet is not a single system but a collection of over 5,000 interconnected computer networks that use a common way of sharing information, Internet protocol as it is called offers numerous basic tools which the public uses in conducting business: sending and receiving electronic mail (e-mail), using Telnet, and anonymous file transfer protocol (ftp). E-mail is arguably the easiest of the three electronic tools to conceptualize because it involves simply sending and receiving messages. The majority of traffic on the information superhighway consists of electronic mail messages because it is cheaper, faster than any other means of communicating a text, extends and enhances the human potential to communicate with others. Other services include electronic publications and e-mail file servers. Geographic barriers are now easy to overcome, as is the formation of professional partnerships. E-mail allows its users to broadcast questions, discuss topics, opinions and transfer documents to thousands of users all over the world at the same time. E-mail is beneficial especially to PR managers who can effectively and consistently communicate to stockholders, consumers, their firm's employees and the general public. Additionally, the Internet had closed the time gap and distance between employees and stockholders. Public relations firms are primary beneficiaries of this technology because they can conveniently access information on any topic. The company only needs coordinating disks to obtain data on any subject. Multimedia computers can enable an organization to make a film or video about its operations, through animation. Some programs target graphic artists, animators, and other creative non programmers who can immediately use their skills to create web pages with interactive media. Managers oversee many aspects of public relations that include maintaining a positive relationship with the public, writing press releases and conducting programs. Because the job of the PR specialist is not limited to these aspects, different equipment and computer programs are necessary. A manager starting a new public relations business should educate his/her employees about the new technologies. They should be able to use computers, cellular phones, and e-mail. They should also be creative and use their artistry on the high-tech systems.

One of the largest attractions of the Internet is the absence of specialized rules; hence both its greatest strength and its greatest weakness lies in its flexibility and its indiscriminate accessibility. Children surf cyberspace and pornography programs, hackers break into secret files using illegal software, researchers extract information from the internet without giving credit to sources, and terrorists use addresses to lure and kill unsuspecting. These atrocities have not been alarming enough to keep users away; rather the many functions of the internet have made it more desirable in "free" societies.

The internet has a lot of available space for new businesses. Automobile, liquor, public relations and other companies have been using CD-ROMS, and other on-line facilities to advertise their merchandise in virtual style. In 1994, no major advertising agency had a group assigned to interactive advertising. However, today most advertising agencies use interactive advertising techniques. The ability to skip from one ad to another can be a blessing and a curse. Unlike TV ads that pop up with no control from the viewer, control of on-line advertising lies in the hands of the consumers. If a consumer does not want to see a company's ad, the consumer simply does not access the company's site. This feature could be regarded as a hurdle to businesses used to advertising their goods via television. The system could be used to see what consumers need. With an organization supplying what is in demand, it can stay longer and prosper. The Internet could also be used for polling interests, services and consumer ideas. One of its largest markets is games. Many families online worry about their children spending endless hours playing video games on the computer. As more educational games are being set up on the web sites, many young people are turning away from books.

With these practices, the employment picture in the twenty-first century is promising, especially if no one reads books. There should be more job openings, especially in the data processing services industry than in any area, as managers move into top positions. In hindsight, over 98,000 jobs were held by PR specialists in 1992 alone. New organizations that turn to media for greater exposure would create more opportunities for public relations experts usually found in large cities where press services and other communications facilities are readily available and where businesses and trade associations are headquartered.

Teletext or Video Text in Africa

Teletext or videotext are subscription services that can be delivered to a residence through a regular TV broadcast signal. The subscriber cannot request specific messages to be sent to his home, but he has a decoder similar to cable TV technology invented by the British in 1974. It is being developed by the French, Japanese and Americans. BBC's (British Broadcasting Corporation's) CE FAX system can transmit a 100-page newspaper or magazine to a home in 24 seconds (Newsom & Wolbert, 1988, p. 172). The teletext could facilitate or expedite literacy and enhance information-sharing and gathering in African and other developing nations. However, only the minority—the very rich and highly educated ones with computers—would benefit from teletext technology. Even the video text delivered to homes through telephone lines, home computers or cable TV systems could only serve about 7 percent of Africa's population. Common consumer services in the U.S. like home banking or shopping could be facilitated in Africa by teletext technology. With free market ideas spreading through developing countries, following the implementation of new capacity-building policies by the World Bank and other international money lending institutions, more local bankers and consumers have been relying on the local banks for investment loans. Thus, they are potential subscribers to the new information technology. However, most of the telecommunication equipment in Africa is outdated and the countries with autocratic and communistic regimes do not have the financial stability to purchase the software for all citizens, hence a capitalist approach to delivering information electronically remains a dream, or does it?

About 40 African countries bought telecommunication services and equipment from AT&T and are developing a partnership with major telecommunication corporations in America and around the world. In 1995, forty African countries were connected to the global undersea fiber optic network which links them to world markets. Regional telecommunication organizations in Africa (RASCOM, PATU,ITU ZAMTEL, INTELCAM, etc.) participated in the project. This wiring of the African continent is expected to yield the following positive results:

1. Government-owned telecommunication departments will own large shares of the profit/income, and
2. African countries may become influential trading partners in the world.

Some African countries are already taking advantage of this postmodern technology to process and disseminate information to more people. About 20 newspaper organizations in Cameroon are using IBM-compatible and Macintosh computers with MS Word, PageMaker and Quark X Press software, to print higher quality newspapers. Mitchell Land (1994) has reported that desktop publishing is providing independent newspapers in Cameroon the means to produce readable material at low cost (p. 43). Land has rightly suggested that U.S. newspapers and their corporate owners donate their desktop publishing equipment and computers to African journalists in order to promote press freedom on the continent (p. 44). This demand could not be made at a more timely period as computer networks are connecting people with different cultures and causing world residents to:

1. Facilitate thinking,
2. Make ever-greater numbers of people in the world to become computer literate,

3. Make people study and use the dominant language of expression in most computers.

The information shipped, some skeptics argue, acculturates and often constructs a condescending attitude in the decoder. To understand the operational underpinnings of this acculturation process and how it influences the telecommunication market, the following segment discusses the philosophical interpretations that underlie communication technology.

Computer Technology and Development in Africa

Because of its ability to store archives of information electronically—that is, in the form of software, text, or image—the computer can meet the needs of any changing society. Computers are a desirable technology for urban areas in Africa. They are being used in government, parastatal agencies, educational settings and the private sector. In the 1960s in Nigeria, third generation computers like the IBM 360 were imported primarily to facilitate government and parastatal activities. The *Service Liaisons d'informatique* in Douala, Cameroon and the Nigerian Ports Authority (NPA), have been storing large volumes of data on import transactions for decades. Other computer systems store and process large-scale data on international educational examinations like the West African Examination Council (WAEC), and the General Certificate of Education (GCE). However, there was a low level of awareness and interest in the computer's impact on social and economic development. According to Okuwoga (1990), head of the Chemical and Allied Products Limited in Apapa, Nigeria, "not so much emphasis was placed on information processing, especially as industries were seemingly being run effectively". Thus "there was far less need for crisis management which necessitates dynamic flow of information for quick decision-making" (p. 102). About 1,600 computers were imported into Nigeria between 1960 and 1989. For a population of over 10 million literate Nigerians to be exposed to less than 2,000 computers in 3 decades may suggest a low interest or awareness of the capabilities of computers in developing that society economically, considering the fact that Nigerians received independence from colonial rule to define and implement their own national development agenda. However, since the 1980s, the importation of computers there has been increasingly rapidly. Over one million computers are in Nigeria today and more computer learning centers are being set up as interest among students has increased.

Although thinkers in industrialized countries may find the current number of computers relatively small for the over 10 million computer-fit Nigerians (those that are educated and can afford to purchase and/or use a computer), its citizens deserve credit for having fore-seen its importance for business, social, and national development and for having become computer literate to the extent that they have. Apart from the numerous institutions in Africa currently teaching Africans to be computer literate, two Nigerian universities have been building computers for Africa-based users. Anambra State University of Technology and the University of Ife, designed and began manufacturing a line of computers in 1983. Hopefully, more Africans will use computers when they become abundantly available for purchase. Moreover, the locally manufactured product is cheaper and user-friendlier than imported ones, and would not require foreign experts and site preparations to be installed. Like the locally designed computers, microcomputers are also likely to be more useful to Third World users than the larger computer systems. They would be used in universities, schools, clinics, and private homes to educate novices. Children would access information from CD ROMS on educational videos and games. Other limitations to computer use might be limited electricity supply, dust, excessive heat and less efficient communication systems. However, support institutions like *CABECA,* Capacity Building for Electronic Communication in Africa have been promoting computer networking in Africa. *CABECA* seeks to provide computer-based networking at affordable costs and accessible to more users. It also plans to make initial site visits, procure, deliver, install and test hardware and software by the year 2000.

Communication Technology: The International Dimension

The arguments advanced thus far have centered on the belief that communication technology in the twentieth century has the power to "globalize" the world. Globalization has assumed new meaning over the years through the actions and interactions of the policy-making elite and nation-states. One of its connotations is "international"—between all nations.

Communication technology increases the degree of interaction between powerful and powerless nation-states and increases economic and interpersonal interactions at the global level, and par excellence, changes the complexion of diplomacy. Through this technology, the past few years have witnessed an increase in corporations, social movements, non-governmental organizations and the emergence of weaker nations. Interactions among states have been augmented by the use of powerful telecommunication machinery. That process has equally increased economic and political indices through tourism, migration and business contacts.

Certainly, the telecommunication industry is changing the way we communicate, it is changing our relationships with each other, our attitudes toward trading, business and governments and our purposes for creating new ones. New microcomputers, satellite dishes, and computer programs are facilitating the creation of new industries around the world. Lindhorst (1991) knows that the advances in technology are commensurate with changing consumer needs (p. 10). According to a report in *The Chronicle of Higher Education,* the U.S. Secretary for Commerce said the American government would develop the Internet into a system for international commerce which would evolve from "'an intriguing communications tool 'into a cybermarketplace' where publishers, writers and other information experts could sell their wares to on-line customers" (DeLoughry, 1995, p. A22). This new attitude adopted by its creators to market communication for profit, and to satisfy consumer needs, shows the importance of having technological innovation in today's society, and of attaining new standards of human taste.

Although inventors and consumers enjoy benefits, it is necessary to note that they have failed to examine cultural implications of globalizing communication technology, especially for nations with relatively small economies. This is a new cultural paradigm that is creating significant unease among international communication schools, ghetto state policy makers, small corporations and traditional development practitioners, in developing countries. Mowlana (1994) has warned that universalizing the practice of communication technology creates an ideological/ economic barrier. In his 1994 thought-provoking article, "Shapes of the Future: International Communication in the Twenty-First Century," Mowlana says:

> Discussions of information highways and cyberspace should be cognizant of the assumption framework of a neo-modernization paradigm. Furthermore, the centrality of the topic of post-industrialization should be kept in perspective. Any discussion of a new world order must take into account the broader ecological communicative context as well as the diversities of a global culture. (p. 26).

The faces and interfaces of culture are so complex and so dynamic that technology, no matter what its sources or modes of production, should not be allowed to erase a culture without considering its practitioner's option. Culture, like a picture, has a huge amount of information and "since every picture we want to communicate may well show very different objects and actions," we must find a method of describing the picture in a way independent of the subject content. (Truxal, 1990, p. 389). The question here is, does telecommunication technology only help human beings worldwide to improve their skills, or does it define the content of all cultures, or both? This puzzle is further explored in Chapter Four.

A Brief History of Information Society

A primitive meaning of the phrase *information society* would be a group of people and digital systems that send, keep, and consume information in a place full of information. An information society does not differ from an agrarian or industrial society in terms of economic interests. Here, information is seen as the core of society's economic needs and a product means more than just goods, energy and services. From a scientific perspective, Lee De Forest, has been credited as pioneer of the information superhighway. In 1906, De Forest invented and patented the thermionic triode valve which triggered an interactive electronic—ideological process—that would become an icon six decades later and would affect people's way of thinking, in spite of their cultural orientations. When De Forest patented the thermionic valve—the basis for the development of television—the American mathematician may have introduced the capitalist world to the era of information technology.

The first technological innovations are known to have appeared during the Paleolithic period, or what some know as the Stone Age—that is, between 2,600,000 BC to 800 AD. During this period, humans learned to make axes, chisels, bows and arrows in order to engage in commercial endeavors. The chisel was used for boring holes in hard objects like wood, and in designing masks, boats and furniture. Bows and arrows were used as weapons to hunt for food, and later during intertribal wars. Later, Stone Age people invented the hoe and other tools for growing crops and learned how to melt tin and copper and mix them. The latter technology was later used in manufacturing printing equipment through which ideas were shared, challenged and expanded. The existence of the Gutenburg Press in the 1400s AD, which according to Hedelman (1992) printed over 8 billion books, and the joint invention of the compound microscope and the telescope in the early 1600s did not only advance reasoning in that century, it also set the pace for a telecommunication invention marathon that went into high gear through the latter part of the twentieth century. The telescope's ability to introduce human beings to satellites in Jupiter and craters in Mars, the invention of the steam boat engine, the telephone and typewriter in 1876, the phonograph a year later and wireless radio in 1895, all gave the Industrial Revolution new meaning. Industrialization could only succeed through telecommunications; hence much effort was made to augment the inventions. The invention of digital computer systems, artificial satellites in 1957, laser in 1960, integrated circuits in 1962, computer chips in 1978, all show an increasing interest in, and need for, using advanced communication technology in shaping a transforming society. The invention of computers, telephones, radio sets and cable only enhanced the idea behind De Forest's creativity. Since 1920, people have been having access to satellites and voice-activated computers. Through such technology today, users can communicate efficiently by simply pressing a button on the computer keyboard. IBM Corporation has developed a computer desktop workstation with a vocabulary of 20,000 words and which can accept spoken commands. Voice-activated computers will be able to digest information and respond to human speech. Certainly, the influence exerted by communications upon society is extensive. Pioneers of communication principles would never have recognized the array of electronic systems that would program minds. Information technology that includes imaging, automation, robotics, sensing technology and mechatronics has affected workers on two levels: it has let workers do their job through an intermediary and it has changed the nature of jobs. Organizations have been obliged to redesign jobs that utilize such technology. A farmer can have the computer analyze soil conditions, plant health and fertilizer, identify individual animals and provide relevant feeding and related health care information using computer bar codes. Even the manager whose primary task has been tailored around time and expertise must now organize, manipulate and disseminate information electronically.

Despite its multiple advantages, information technology has not suppressed the suspicion of users who fear that their unprotected information can be proliferated, stolen or misused. The right to privacy is valued by people in all cultures and mankind has sought to conceal certain information for the purposes of maintaining order or integrity, and hence for strengthening its ego. To maintain human rights, governments in free societies have been seeking policies that oversee the protection of the privacy of electronic information users, although such policies are still limited. The only statute in the U.S. that addresses this issue is the 1986 Electronic Communications Privacy Act (ECPA). ECPA prohibits the interception of e-mail messages by entities outside a company except proper legal authority is given (Cappel, 1995). Because this law only monitors the activities of external organizations, parties within a company are still free to intercept messages. The ambiguity of this law is, employers may monitor and read their employees' e-mail transactions and vice versa. The international implications of Internet privacy are further explicated in Chapter Five.

Exporting Technology

The value of the composition, organizational structure and quantity of technology transferred from a country is determined by the recipient country's level of economic development and industrial capability. The establishment and licensing of intellectual property rights—trademarks, patents, and copyrights—require technical and managerial assistance. This is where entities in the recipient countries lack expertise in translating documents from the exporter for their own benefit. Licensing agreements in African countries have often constituted a joint venture between domestic business executives (government officials) and foreign technology suppliers. As discussed later in this book, this is a practice which promotes bigotry and self-centeredness and does not consider the interests of the majority of people. Nor does the technology cause a significant change to the people's economic status.

The exportation of technology to developing countries can be appropriate only when the technology causes psychosocial and infrastructure change. There are two levels of technology that require explanation: revolutionary and non-revolutionary. Revolutionary technology refers to knowledge and/or systems, which provide an opportunity for innovation. Here, knowledge has to be specific and it must have a defined quantity in order to generate greater knowledge or change. Non-revolutionary technology means knowledge or systems, which only permit the operation of routine tasks using specific equipment. A fax machine does not generate knowledge, it only transmits information. Since the end of the Second World War, industries have been focusing on the significance and creation of value-oriented technology. Technology is useless unless its function is predetermined. The use of a technology should be determined before a decision is made to import it.

For technology to be exported there has to be an agreement on implementing certain codes of conduct by the parties. The Intergovernmental Group of Experts prepared a draft code aimed at establishing standards for the transfer and transaction of technology. According to Roffe (1985), the draft code sought to:

a. determine which practices involving technology transfer should be considered undesirable and under what conditions,
b. identify and clarify the rights and practices to transfer technology, c. provide laws governing technology transfer and evaluate the extent to which parties might be free to choose a regulation and opportunity to settle transaction disputes. Roffe (1985) added that the code sought to establish standards applicable to all groups and all countries involved in the transfer. This issue is further discussed in Chapter Six: Telecommunication Policies in Africa.

The exigencies of global competition, capital accumulation, and the desire for profit-making have expedited the capital intensity of technology. Seyoum (1990) has provided reasons for the growing number of industrial jobs in developed countries. He has said the number has been generated by increasing volumes of production and numbers of industrial establishments. Seyoum (1990) has further cautioned that undue emphasis on labor intensity may cause reduced quality and efficiency and may increase the cost of production, thus putting the industry in a position of competitive disadvantage in international markets (p. 40). Based on that argument, the performance or input of a given technology should not be attributed only to its ability to generate employment. Prior to its importation, such factors as product quality, product choice, product demand, law, and the political climate in the importing country should be seriously considered.

Technology, Government, and Law

Technology, government, and regulations complement each other and control the future of the free society. Without laws there is chaos, without technology there is poverty. The relationship of the triad provides a climate conducive for the peaceful coexistence of government and industry. For example, the U.S. model for industry-driven regulation shows a balance of power between the government (regulator), the provider (hard/software telecommunication industries) and the user. The government is the regulator, industries the provider, and the consumers the users (see Figure 1.2).

This model is based on the concept that the consumer, seller and the lawmaker have an equal amount of influence on each other. The model is not yet applicable to developing countries because providers of hardware and software technologies operate primarily from abroad, without considering local users' input, and corruption among government officials weakens the implementation of telecommunication laws. It can only be applicable there if governments in developing countries decide to expand their economies.

To articulate the role played by transferred communication technology in the political and economic futures of such governments, the next chapter describes the products and activities of selected international markets.

Figure 1.2
U.S. Model for Industry-driven Regulation

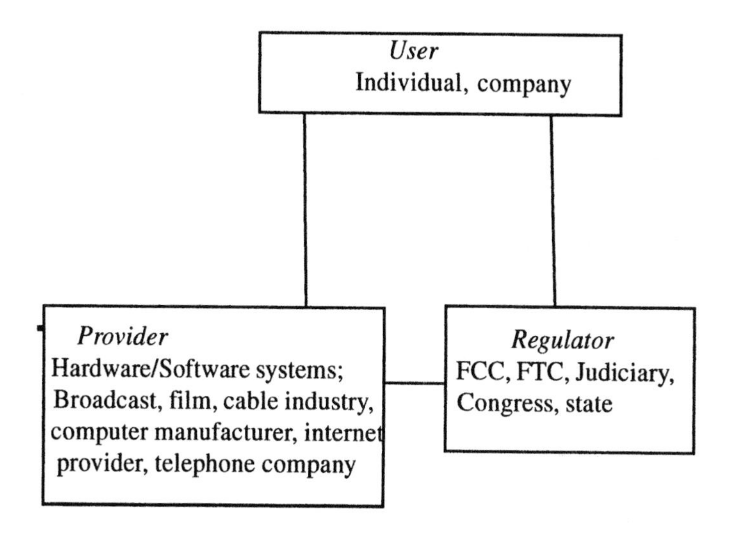

Part II

Infotech and Socioeconomic Development: A Market Overview

Chapter Two

Foreign Telecommunication Companies: Market Dynamics and Services

Overview

The potentials of foreign telecommunication markets in establishing business in developing countries can be best examined in terms of product value and product availability. This chapter describes the services telecommunication companies render domestically as well as the politics surrounding their operation. The description of their domestic activities is meant to determine their potential in operating in foreign markets. The chapter further describes the nature of competition among U.S. long-distance telephone companies, based on the following concepts; the telephone is the fastest growing product in the industry, the most influential agent of communication, and the most affordable product, and can generate greater business in Third World markets which have limited financial resources. The chapter also provides the social impact of telephone on both Western and Third World societies.

Cellular Phone Business: U.S. Connections with Canada, Japan, Mexico, and China

Since the 1950s, America has had an almost constant flow of communication inventions infiltrating day to day living. The remote control that almost every home has and cannot imagine living without, or the beloved cordless phone that enables people to do a hundred other things while on it come at a time when our sense of inquiry is more daring. In the pages ahead, the discussion will center on the cellular phone, a piece of high-tech equipment that can connect a telephone call all over the world.

Until 1990, mobile radio was the wireless communication agent for intelligence agencies and business organizations. What used to be the industry for public servants is now invading private and corporate sectors in America. Motorola, the pioneer of cellular service, created the device on a gamble, but then reported a billion dollars a year in business in 1994. Cellular service has almost tripled, to approximately thirteen million subscribers in only five years. However, some experts believe this technology is beginning to lose ground. (*Time,* November 1994, p. 38) reported that 40 percent of cellular phone calls in high density areas such as Manhattan, New York and Los Angeles are not complete. In 1994, when a Florida man filed a lawsuit claiming his wife contracted a fatal brain cancer from heavy cellular phone use, it appeared that one of the 1980s biggest, fastest-growing industries was headed for trouble. The cellular phone industry has blossomed into a global business, with such countries as Mexico and Japan becoming major users. The scare had an effect, because the largest provider of cellular phones, Motorola, which had expected its sales to grow by 40 percent, lost sales.

But the expansion was not without its share of political and economic traps. In December 1994, trade negotiations between the U.S. and Japan stalled when Japan refused to offer concessions that would help relieve the trade deficit. In retaliation, the U.S. threatened trade sanctions unless Japan granted access to its cellular phone market . The Japan-U.S. debacle would never end as long-as both capitalist nations produce goods and compete for the same markets.

Canada

Canada's case is different. When Canada, with a $6 billion long-distance market, opened its doors for long-distance competition, it allowed

AT&T, Sprint, and MCI to offer "superior services and cheaper rates" in the U.S. (Symonds, 1992, p. 36). This move basically captured the Canadian clientele. Canadian-owned United Communication Inc. allowed the services of U.S. companies because Canadians were paying 60 percent more on long-distance rates at peak time, which handicapped their business community in global competition.

Japan

The Nippon Telegraph and Telephone Corporation's installation of a $400 billion optical network expected to be completed by the year 2010 in Japan will make the country's information technology and cable market highly competitive. The Japanese Telecommunications Council projects that a fiber optic network would serve 75 million subscribers and would cost $800 billion including underground cable. This would increase the percentage of Japanese cable subscribers from 3 percent to over 60 percent, smaller than the current figures for U.S. cabled households. The Japanese Telecommunications Council has been reviewing laws that restrict the overlap between telecommunications and cable industries. That review of telecommunication regulations indicates Japan's interest in modifying its culture to cope with the New World Information and Communication Order (NWICO) that itself promises fusion of world cultures, politics, power and the sharing of economic profit. Although Japan is a leader in electronic goods, the cellular phone business has not yet saturated the Japanese market because of embargoes. In 1995, United States effort to force Japan to open its cellular market to U.S. competitors was regrettable for many cellular distributors could lose millions of dollars from the world's largest trading superpower. The United States might have difficulties controlling the lucrative African market because Japan has a long-tradition of exporting its products at low and affordable rates to Africa. Motorola can do more business in Mexico and other developing countries in the western hemisphere because of proximity and the terms of the North American Free Trade Agreement (NAFTA).

Mexico

Mexico is not half as technologically advanced as Japan or the US. In Mexico where poverty abounds at an amazing 52 percent rate, a cellular phone takes less precedence to homelessness, hunger, and disease. When NAFTA was passed allowing the United States and Mexico to trade without tariffs, it enabled the cellular phone industry to have less difficulty marketing its product than in past years. The arrival of this new product would affect Mexicans socially, educationally, and economically. By allowing free imports, prices will be cheaper than in the United States, and working class Mexicans will be able to purchase this equipment, hence increasing the potential for marketing their skills beyond national borders. Although unemployed Mexicans might feel resented for being unable to purchase this equipment most Mexicans will gain from NAFTA because more middle and low income jobs will be created. Telefonos de Mexico, commonly known as Telmex, has upgraded its equipment and has expanded services to remote corners of the country, in order to prepare itself for foreign competition, when the government opens the long-distance market for the first time. The new competition is expected to lower service rates for consumers and reduce Telmex's revenues. Telmex is a wise investment because of the growing demand for telecommunications, engineered by the NAFTA in particular, and the country's growth potential in general. Also, the company has proven its effectiveness in improving local service and will keep its local service monopoly until 2026. The July 18, 1994 edition of the *New York Times* reported that the attractive Mexican market caused MCI, Motorola and Bell Atlantic to form separate partnerships with Mexican banks and telecommunication companies, while AT&T and Telmex negotiated the formation of a long-distance subsidiary.

The latest technological breakthroughs could also open a whole new world of experiences for students and telecommunication trainees. Instead of hearing speeches or seeing diagrams of fiber optics, and electrocells—a common practice in developing countries—, trainees would have the opportunity to use the objects they usually theorize about. In addition to fax machines, cellular phones would help to introduce the students and trainees to this ever-changing world of technology.

Mexico has tried unsuccessfully in the past to revitalize its communications industry with Nextel Communications. Nextel's new alliance with Mexican Cellular Telecommunications firms follows an agreement between Nextel and Canadian based Cleanat Communications Inc. for Nextel to develop a cellular communications network that would enable cellular phone customers to use hand-held equipment to access cellular services throughout North America by 1996. This technology will allow the company to reach about 75 percent of Mexican residents in both urban and rural areas. These inventions are only a catalyst to other major inventions like the Iridium, which in 1996 consisted of 66 satellites orbiting about 500 miles above the earth. The satellites will put together a network of special pocket phones, personal computers and pagers anywhere in the world, for a subscription rate of only $3.00 per minute. Such devices will also be beneficial for everyone if introduced soon.

China

China's limited use of telecommunication equipment should not surprise anyone for several reasons: its economic and political situations make China a Third World country. Ranked below 100 countries in per capita GNP, with only 10 percent industry employees, and a large population with an economy controlled by a communist government, one may not expect eagerness among the Chinese to have an effective telephone service. A monthly residential bill costs two year's salary in China, so most people must rely on public phones. In the early 1990s, China had only two telephone lines per 100 people and by the year 2020 there shall be 40 people per 100 lines. In 1993, AT&T arranged to build phone switches and planned with the Chinese government to set up a research Bell laboratory in China, to build switches and cellular phone systems and to show the Chinese how to make microelectronics for telephones. As the Chinese attempt a free market economy (think of Tienanman Square demonstrations), and with a 60 percent investment in Chinese fiber optic-cable plan, AT&T could yield substantial profit as long as the corporation obtains a waiver to an international agreement that prohibits the exportation of high-tech equipment to the communist country.

Telephony and the European Market

In 1991, Anglo-Dutch company, Unilever PLC signed a three-year contract to run its 17-country European Communications network; computer data, electronic mail, and voice. Later, British Telecommunications PLC or Royal PTT Netherlands was outclassed by U.S. based Sprint corporation which promised lower rates, guaranteed quality service, and a large network to choose from. Additionally, the European Commission (EC) thought the bill for telecommunication services in Europe would rise from 3 percent of Europe's gross domestic product in 1991 to about 7 percent by 2000 AD., following EC's 1987 policy to abolish monopolies in data transmission and electronic mail among its 12 member states. The globalization of communication has become a reality, especially to industrialized European countries which already had more sophisticated infrastructures compared to Third World countries. EC's intention to open satellite, mobile communication, and private data networks to other competition only asserts the former argument. So far, the telephone business is the foremost beneficiary of the demonopolization process because telephone calls made across national borders, especially between European countries, are cheaper and more reliable than other data transmission systems. Although European countries expect to yield a 90 percent profit in international long-distance calling business, U.S. companies have a better opportunity to control the long-distance market in the future because they offer lower rates, and more efficient services. They use more sophisticated equipment, and provide new jobs to economically and technologically limited regions like Africa and the Middle East. In order to understand the level of competition among U.S. long-distance phone companies, an understanding of the history of the telephone is necessary.

Telephone History

The origin of the telegraph system in America can be traced back to 1876 when Alexander Graham Bell invented the telephone. This was followed 4 years later by the creation of the American Bell Telephone Company. The commercial long-distance line opened in 1881 and in 1989 American Bell was taken over by AT&T and was given a telecommunication monopoly by the United States Government. AT&T was responsible for providing local services through 20 bell telephone subsidiaries for an estimated 85 percent of the population and long-distance services for about 100 percent of the population until the 1970s. According to the 1993 Telegroup Representative Manual, there were no separate services available for customers. AT&T maintained monopoly over long-distance services and implemented high rates and a three minute minimum billing system called Messages Telecommunications Service (MTS). By 1974, AT&T owned a $150 billion network, thereby over powering every aspect of the industry from switch and telephone manufacturing, to local distance service. Since there was no legitimate competitor, subscription was monopolized by AT&T and thus, subscribers remained vulnerable to any incompetence in the industry.

Telephone companies like MCI, Sprint, ITT, and AT&T have been rendering private line services to large corporations and residential areas for decades. MCI, formally known as Microwave Communications, Incorporated, built its network using microwave tower to improve the quality of its services. In the beginning, its target market were two large companies with network sites and large amounts of traffic between major metropolitan areas. In the early 1970s, MCI sought to connect its network with the Bell Telephone System. Threatened about a new competition, AT&T rejected MCI's bid to make the connection. In turn, MCI sued AT&T for three billion dollars in damage and for the right to make the connections. The issue ended up with U.S. Court Judge Green ordering AT&T to break up. As a result, AT&T became a long-distance company above seven local telephone companies. AT&T's monopoly in controlling local services ceased, while Regional Bell Operating Companies were given the right to offer just local services. MCI acquired one billion dollars as compensation, while businesses and residential areas got lower long-distance rates. The rates have not been quoted in the Telegroup document and remain unavailable. However, MCI and Sprint built their own nationwide networks and started competing successfully with AT&T in almost all service areas.

The emergence of MCI, Western Union, ITT, and Sprint in the early 1970s forced AT&T to lower priced services. These new companies constructed their own private networks and provided their first service called WATS-Wide Area Telecommunications Services. Although this service created major problems for the customer by providing expensive installation, charging monthly fees and installing inflexible lines, these companies offered better rates than AT&T , thus making WATS a higher competitor to AT&T. To combat this and to develop a new market, AT&T created the PRO family program in the 1980s. AT&T, Sprint, and MCI operate in a capitalistic society where the struggle to manipulate prospects and make profit is lifeblood of corporations. Whenever there is tight competition, some competitors go out of business because they lack enough customers, or capital, and because they undersell their products, or do not have quality products to stay in business. Companies use different advertisement strategies to attract and maintain customers and to generate more interest. In order to understand the level of competition that exists in that industry, we need to analyze the nature of consumer interest and market strategies applied by the companies, from a mass communication perspective. However, some scholars do not regard the telephone as a mass communication medium. Teleconferencing is indubitably a mass communication activity because many people speak or listen to each other at a given time using different telephone lines. The caller-connect/caller-hold system enhances teleconferencing through the two-step flow; corporate managers and group leaders spread their decisions among their companies and loyals after a teleconference.

Market Size

In any free market enterprise, the consumer is the "king" any advertiser strives to satisfy the satisfy, because advertisers depend on the former's purchasing ability to function as business units. Hence, there is usually a high level of competition between competitors to offer the best services or products in order to maintain consumer interest and to attract more consumers thus maximizing profit.

There are three major types of consumers: the skeptical consumer, the optimistic consumer and the free-spending consumer. The skeptical consumer, usually a low income person, has limited funds and requires a lot of persuasion from the advertiser, in order to purchase a product. That consumer has to be constantly reminded or exposed to subliminal advertisement of the service or product on TV, newspaper, magazines and other mass media before he can purchase the product or service. Advertisement brochures mailed to him including follow-ups, may persuade the skeptical consumer. The optimistic consumer does not require much persuasion to purchase a product. That consumer, usually a person with a fixed budget, can easily buy a product or sign up for a service whenever he has bargaining information. He makes a choice base on what best suits his needs or whenever he is exposed to an advertisement. The free-spending consumer, unlike the skeptical or optimistic consumer purchases a product as soon as he is exposed to its advertisement. That consumer may purchase a product after becoming exposed to it through the mass media through or an individual. As long as he has the money and needs the product, he purchases it right away.

These three categories make up the market for long-distance telephone companies. AT&T, Sprint & MCI developed sophisticated strate-

Table 2.0: *NTC, AT&T Sprint Pricing~4.5 minute residential day time call versus NTC saving*

Bands	Mileage	NTC	AT&T	Sprint	MCI
US1	1-10	0.8487	1.1500	1.1000	1.0995
US2	11-22	0.8487	1.1500	1.1500	1.1475
US3	23-55	0.8858	1.2000	1.1500	1.1475
US4	56-100	0.9225	1.2500	1.1500	1.1475
US5	100-124	0.9225	1.2500	1.1500	1.1475
US6	125-292	1.9225	1.2500	1.2000	1.1975
US7	93-430	0.9594	1.3000	1.2000	1.1975
US8	431-925	0.9963	1.3500	1.2000	1.1975
US9	926-1910	0.9963	1.3500	1.2500	1.2445
USl0	1911-3000	0.9963	1.3500	1.2500	1.2445
US11	3001-4250	1.1070	1.5000	1.5000	1.4995
US12	4251+	1.2177	16.500	16.500	1.6495

Day rates based on a $25 long distance bill, with 18% prompt pay discount NTC has a 6-second billing. All other carriers have 1 minute billing system tariffs available on January12, 1994. Many price changes are expected given the ongoing competition among carriers This is a modification of NTC's Form 1213.

Table 2.1: Evolution of major US Long Distance Carriers
in market shares and estimated profit

Carrier	1987		1992	
	$ billion	% share	$billion	% share
AT&T	34.4	79.9	35.5	36.0
MCI	39.9	9.0	8.9	16.0
Sprint	2.7	6.0	6.5	11.0
Other IXers	2.6	6.0	4.2	8.0
Switches IXs	0.0	0.0	1.0	2.0

gies to entice consumers and made them feel they were getting the best deals. However, in recent years, the market percentage share and profit have differed dramatically between these companies, (see Table 2.0).

To give each company an equal opportunity to assist in the transmission of long-distance services to full competition, the Federal Communications Commission (FCC) adopted a price cap regulatory framework in 1989, which included: Baskets One, Two and Three telephone services. Basket One telephone service consisted of a calling card and traditional long-distance and international calling service, which were purchased basically by residential and small business customers. Optional calling plans also offered discounted and residential long-distance service for a charge, based on how a consumer used the service. Among these optional calling plans were AT&T's "Reach Out America," MCI's "Friends and Family," NTC's "Half Price Club" and Sprint's "The Most World Wide." Basket Two and Three services included "800" services and private lines that were purchased primarily by large businesses. Most of the Basket Two and Three services had been shifted to streamline regulation in 1991 and 1993. Federal Trade Commission (FTC) staff economist, Michael Ward suggested that the market for Basket One service which was thought to be less competitive than Baskets Two and

1991 data on AT&T, Sprint and MCI (FTC Report, October 27, 1993). The study also showed that optional calling plans offered discount rates to high volume callers, that is, callers who had large monthly phone bills. Certainly, these customers had more to gain from the calling plan when they shopped for lower prices. The estimated shares and profits from 1992-1995 project the industry as having the potential of becoming a highly lucrative marketplace for investment (see Chart 2.0).

Following the decision by the U.S. Defense Department to make the Internet available to the public, the 1996 Telecommunications Act, (which promised greater availability of equipment and services), and the capitalist principles upon which the American character is built, there is no doubt that the telegraph market is expanding more rapidly in the 1990s. Over 90 percent of U.S. residents own a telephone set and over 90 percent use one form of telecommunication appliance. According to the Telegroup Representative Manual (1993), an estimated sixty billion dollars is spent each year on long-distance services for residential and business customers in the US. On the other hand, only about fifty billion

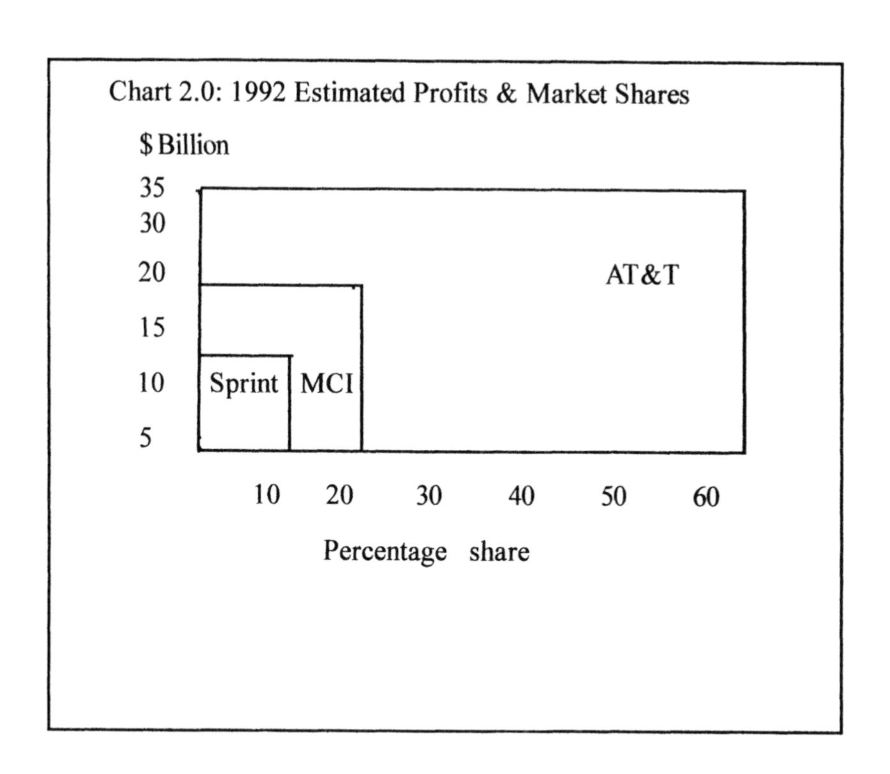

ness customers in the US. On the other hand, only about fifty billion dollars is spent on long-distance services in the rest of the world, hence there is ample need for serious competition among long-distance telephone companies to attract more customers abroad. In the provocative paper, "Mobile Satellite Communications: From Obscurity to Overkill," Whalen (1995) has predicted that over 38 PTT companies will invest in the telephone market and they will yield $10 billion in annual revenues for 11 million subscribers at under $1 per minute call. The question arises: Can the long-distance telephone market support all upcoming companies? Would new and more sophisticated computer systems like the Internet and E-mail beat the telephone market worldwide and sustain individual privacy. Already, the Japanese have created a telephone system that reveals the caller's face on the receiver's phone. Caller IDs and speaker phones suggest major earthquakes in the long-distance communication industry. These advancements no longer guarantee individual privacy, although caller IDs and caller pictures might reduce the rate of "crank" phone calls and increase consumer privileges and rights to privacy.

The economics of telecommunications shows a steady rise in the United States. Most white-collar and blue-collar workers prefer to use cellular phones and beepers because they are portable and convenient. Business people do not have to stay home or in the office to return phone calls. The cellular phone industry has over 12 million subscribers. Two-thirds of them are in Japan and the U.S. So far, America's telecommunications services and hardware providers are competing with rivals like Canada's Northern Telecommunication, the Japanese Nippon Telegraphy, Siemens of Germany, and French Alcatel. Highly developed and relatively open telecommunications networks in the U.S. offer the most tempting opportunities and foreign firms are receiving shares of the equipment market (*U.S. News and World Report,* 1990, p. 44). The U.S. has the largest and most advanced telecommunications system in the world, followed by Japan, Germany, Britain and France. Chile and Argentina have also opened up to competition. On the other hand, France does not have much experience in changing from monopoly to competition, given its political ideology. The government implements a dirigist approach. So business incentives are generally geared toward protecting state security, rather than toward establishing economic plurality and opening up markets for more competition. For example, the strategy for the information society designed by the French government council on October

27, 1994 did not have clear and quantitative objectives, or a binding schedule with public telecommunication operators. Instead, it advocated private enterprise on a trials basis (Vedel, 1996). It is easier for the French government to negotiate with parties and secure business operations than to allow companies to create new markets. For French products to be as marketable as those of other industrialized nations, France needs to seek more private markets in the Third World improve the quality of its products of its products and lower its prices. Telecommunications management is not a priority for Third World governments mainly because it does not provide short-term benefits. Telecommunications could be useful to the Third World by enabling people to communicate important messages faster within their own villages, nationally and abroad. On the other hand, these countries would have to provide better services to international callers in order to be competitive with the more technologically advanced countries. Third World countries are new to this industry, so the outcome would be no contest.

Marketing Policies

Telecommunication policies are now a complex mixture of robust competition for long-distance telephone service and for exclusive franchises for cable operators. President Clinton and Vice-president Gore have been committed to removing many of the legal and regulatory problems that prevent the emergence of a competitive information superhighway. In *U.S. News and World Report* (1994), U.S. Vice-President (1992-2000) said the industry could "promote and protect competition, encourage private investment, provide open access to the network, and protect the long-standing policy" (p. 56).

The U.S. government has showed much attention to telecommunications by implementing policies that may delimit monopoly among other telecommunication networks. The 1996 Telecommunications Act gave providers more flexibility for free enterprise. Providers could market their products in any market providing they maintained trade laws. The terms of this and other agreements are also examined in Chapter Six. With the government trying to achieve these widely supported objectives in the face of fast-moving and competing interests, there would be unprecedented developments. For example, MCI, the nation's second largest long-distance company declared that it was invading the monopoly turf of the big regional telephone companies. With a $2 billion investment, MCI was able to connect businesses directly to its long-distance network from some twenty U.S. cities.

The deregulation of the telecommunication industry has helped companies to manufacture more quality products and has allowed long-distance companies to compete with local providers. These companies have been actively utilizing aggressive campaign strategies to cultivate, maintain, and increase their markets. For many years, they have used direct mail, flyers, broadcast media, newspaper ads, telemarketing and person-to-person contact. Some telephone powerhouses have been bought while others have merged to heighten competition and generate more profit. On November 2, 1996, British Telecommunication (PLC) attempted to purchase MCI Communications Corporation for $20 billion, which would have made British Telecommunication a legitimate competitor with AT&T Corporation. Although MCI would no longer be an independent company, and FCC stipulates that foreign companies cannot have indirect ownership of more than 25 percent of a U.S. company holding a wireless license unless the deal promotes public interest, U.S. and foreign clients previously using British Telecommunication and MCI services would have cheaper prices and better services.

AT&T's merge with McGraw Cellular Communications, Inc. also shows its vision and plans for the future. In an effort to make communications possible anytime and anywhere, AT&T's high tech communication systems enable a telephone call or fax to reach customers any where and at anytime, even when sitting in a plane or resting on a beach. According to *Business Week* (August, 1993) AT&T's plans included voice electronic mail and video conferencing. These services zapped across AT&T's digital information superhighway are delivered to conventional phones and computers, television's cellular hand-sets, and to a new generation of wireless devices. AT&T has had fast- moving, all digital multimedia with a major ability to conquer any market. Sprint has also acquired wireless system after merging with Centel and, like MCI, which sold its 20 percent stake to British Telecommunication, Sprint has sought to integrate the Centel Cellular service with its own wire network. Its partnership with French and Dutch telecommunication companies will ultimately increase its clientele.

The Telephone as a New Cultural Paradigm

Americans are very familiar with electronic communication and would not be able to function normally without using it. Telecommunication products have become a paradigm of the American culture. Businesses and major corporations rely so heavily on beepers, cellular phones, fax machines and computer inter-communications systems that most of them could go out of business without this technology. Americans cannot survive without new forms of telecommunication technology. Not only do companies depend on the telecommunication network, homes and families do too. Today's baby-boomers live such fast-paced lives that they need to call back home to check on their families. In the corporate world, the technology allows employees to relate messages with just a push of a button. This process diminishes chances of missing business meetings and increases opportunities for calling back important clients on time. Domestically, having good telecommunication equipment allows busy parents to leave messages for their children while away at work. Also, an answering machine records messages while the individual is away from home, and lastly, people can page their loved ones anywhere when lonely or in trouble. Car or cellular phones have become America's latest security guard. Whether a subscriber is involved in an accident, a robbery, an act of violence or a decapitated vehicle, a rescue squad is summoned to the scene in a matter of seconds with a cellular phone call. For example, Sprint Corporation's Cellular One plan includes services for motorists experiencing a breakdown. The Cellular One subscriber is eligible to receive car fuel on the highway when it runs out.

Of course, there are disadvantages in using high-tech equipment. Americans tend to let go of traditional values such as family togetherness. Reading together or chit-chatting as a family has been replaced by electronic individualism. Television, radio, phones, and fax machines now absorb the time that families should invest in each other. Americans may be obsessed with and abusive toward modern telecommunication technology because it is readily accessible to them, whereas for Third World prospects, it is a scarce commodity. It would give the latter a different perspective on how to communicate with one another. Also, innovations in electronic media like high definition television would allow them to learn more about the world, their personal environment, and different cultures within a shorter time. However, culturally, many Third World countries have strong histories and family backgrounds. Residents make their own clothing, fetch their own food, and speak their own language. Bringing such technological advancements to these countries could diminish their values as it has already done in American culture.

Given the status quo, where will new telecommunication technology go from here? Will it become a greater economic phenomenon? Will it be taken over by governments and corporations? Or, will it destroy cultures and traditions ancestors worked hard and preserved for generations to sustain? Can Africans benefit from the information supermarket without having new technological infrastructure? Can they successfully exploit new technology without being exploited by it? How can Africans use the Internet without understanding the concept of having cheaper commodities like telephones? Against the backlash of these questions lies a new perspective toward the establishment of significant changes in Africa. As government authorities and scholars grapple with those issues, the countries harboring AT&T infrastructure and services are undergoing their fourth wave of telecommunication technology. Most middle class and third generation Americans and business entities use pagers and other portable cyberspace technology, while most Africans are only becoming acquainted with color television and telephone use. Even the so-called African One project sponsored by AT&T Corporation to install a cable system in West African coastal regions and which planned to provide quality electronic communication services to Africans is a suspect endeavor, a contemporary geographic picture reminiscent of the export route used during the international slave trade movement. Arguably, it is another ploy by powerful telecommunication companies to exploit Africans, given that African Posts and Telecommunication Departments anticipate only a small percentage of all financial benefits. Moreover, Africa One is only a 20-year trial venture, so by the time Africans begin to yield economic benefits from the project, AT&T Corporation would be pulling out.

Here, one can see a pattern consistent with that executed by foreign slave buyers. Like the buyer who used local interpreters to coerce local leaders with intoxicating drinks and clothing to negotiate the release of their people, the telephone company CEOs have arranged with respective officials in African governments to install fiber optics systems on African soil without consulting with tribe group leaders to whom most

Africans are loyal and whose ancestral land they are using. Local residents have not been asked whether they want these services or whether they understand the concepts of using foreign telecommunication technology.

Since the primary function of a company is to make profit, AT&T's Africa One fiber optic project may not include the hinterland whose per capita income is the lowest on the continent. Hence, the residents would not afford the cost of AT&T's services. Additionally, telecommunication users in industrialized nations barely manage to pay their bills, hence it is illogical for any company to bill African rural residents who rank among the poorest class in the world. AT&T's construction of the fiber optic network only along the coastal regions is meant to facilitate business transactions between Africa and Europe, especially on oil, money and raw materials in Africa which are in high demand in Europe. A keener look at Africa One network shows easy access for AT&T to transact business between Africa, France, Germany, and the Middle East. Also, it would cost AT&T Corporation more money to install the infrastructure because of geographic obstacles: rivers, lakes, mountains and primitive natives capable of destroying the electrical wires and sabotaging technicians' work. Projects of this nature have been suspended in Senegal and other countries because rural residents reportedly dug out cable wires to design earrings and other amulets. Whether these residents were aware of the original usefulness of this equipment to the advancement of their economic status, they would not likely benefit from the information networking so promised by AT&T executives because they do not have economic value in the eyes of these companies. Moreover, they are uneducated, poor, have limited knowledge of Western culture (the 'cradle" of modern technology), and they manufacture products with limited international market value. Hence, distributing fiber optic network services to the rural areas would be a waste of resources for any profit oriented organization. Even the United Nations which in 1971 advocated the installation of telephone lines in all rural areas has not fulfilled its request. The only practical solution to these problems is for the respective governments and AT&T Corporation to mediate with development communication practitioners and local authorities in order to determine a market size for the corporation's products. The corporation can also train and employ local residents. Installers should avoid sacred grounds. Providers and government authorities should confer with local authorities prior to exploiting local space for equipment installation.

Products and Services in the 1990s

As earlier mentioned, competition between the three major long-distance companies keeps the companies busy with innovative techniques, including an offer of various discount rates and coupons to lure and maintain local and international customers. AT&T's "I" Plan, an individual account savings plan included "Reachout America," "True World Savings," and ' Reach out World" discounts, could be accessed by calling toll free numbers. Rates for the "Reachout America" discount plan have differed depending on the caller's location and state/province he/she was calling to, as well as the time the call was made. According to customer representatives for the AT&T office in San Antonio, Texas, (September, 1996) AT&T had three long-distance rates; the Standard, Economy and Evening plan. The Standard rate has been the most expensive, the evening rate cheaper than the standard rate, and the economy rate the cheapest. The economy rate for AT&T was about 40-60 cents cheaper than the standard rate. The economy rate applied from 11 p. m. to 8 a.m. and calls made during this time cost 13 cents per minute. Under the economy rate, a customer could save money only if the call was made within the continental U.S. Calls would be measured in terms of mileage. Thus, the further away the receiver from the caller, the higher the amount of money to be paid at a given time. The evening rate cost approximately 6 cents more than the economy rate. This rate became effective from 5 p.m. to 11 p. m. on weekdays. The standard rate operated at approximately 25 cents per minute if a call was made within the U.S. It also depended on the distance between the caller and the receiver. Hence, the further the caller, the more expensive the call.

As far as international calls, AT&T's "I" plan is complicated and long-distance calls differ from country to country. The number of calls received in some foreign countries depends on an agreement reached between that country's telecommunication officials and the AT&T company, meaning the rate would depend on the number of phone calls the country allows into its lines. A more detailed analysis comes later in this chapter.

Since certain countries have few lines, circuits were usually busy when many people call into the few lines. On a public holiday in Africa like Independence Day and Christmas Day, telephone lines got jammed up with phone calls from citizens of those countries living in the U.S.

Except in the mornings and nights, calls frequently made were not clear because of static and other connection problems. It became difficult for the customers to enjoy the privileges of the long-distance discount rates being offered by their respective long-distance carriers. Sprint and MCI have , however, alleviated this problem. Since 1991, calls to most African countries have been transmitted smoothly because most of the phone lines and busy circuits were cleared. Also with AT&T's installation of fiber optics in Africa, telephone transactions will be easy.

Although some international calls go through with limited problems, companies have not been able to satisfy most residential long-distance callers because of the ever high rates. For example, a phone call to Japan during standard time, that is, between 8 a.m. and 5 p.m. weekdays on AT&T costs approximately $1.05 per minute. Although there is a time difference between the U.S. and Japan, the long-distance telephone companies still fail to reduce the telephone rates. It must be noted that standard long-distance economy rates differ from country to country. For example, if the caller who made a prime time call to Japan called Cameroon, the call would cost him $2.50. MCI subscribers for international calls to Japan, or Cameroon are expected to save money, especially when they call during off peak (non-prime time) or weekend hours. For instance, based on a promotional program for 1998-99 fiscal year, a regular phone call to any city in Cameroon costs $1.04 (approx. 600 francs, CFA) a minute, but calls lasting 2 or more minutes are charged a higher rate. Most telephone companies apply a first minute surcharge of about $1.90, except Unidial Communications which only charges on a 6-second increment basis.

But the surest thing about telephone rates is that they change constantly. The telephone is the fastest growing business in America in the 1990s with a 1993 $185 billion dollar income. By the year 2000, long-distance telephone service will be a $300 billion industry. In 1998, the sales growth of long-distance products was over $90 million. This growth was the god-child of such products and services as Dial-I Residential and Business Service with a 12 percent–46 percent saving plan. The Half Price Club mentioned earlier provides a 50 percent discount for the second year and 60 percent the third year, all with a digital fiber optic quality call transmission and free travel incentives. Like the other three companies, NTC encourages customers already using another long-distance carrier to switch over to its services. It usually costs a reimbursable $5 to switch to a different carrier. The competitive rates shown in Table 2.1 explain why the long-distance industry will become one of the most powerful profit-making industries in the 21st century.

Most countries are being wired through telephone lines to facilitate long-distance and international transactions, hence the companies stand a stronger chance of reaping profit than ever before. Because of the competition between these companies, rates and calling plans change constantly. The telephone companies have further realized the need to advertise their products including calling plans like, Half Price Club, and New Friends & Family, to customers in the U.S. and around the world. Their persuasive strategies have ranged from promising customers free trips and using retail outlets like fuel stations for selling prepaid phone cards, to offering free telephone calls. MCI has been attracting its own worldwide clientele, through such products as the Calling Card. World Phone and Fon Card. (See Table 2.2).

Prepaid Calling Cards (PPCCs)

In Abidjan, West Africa, PPCCs are used when calling within the city at 33 cents per six minutes. Ironically, calls from one African country to another are more expensive than intercontinetal ones.

Table 2.2: Convenient Call Placing on MCI WorldPhone Plan for Foreign Callers to the US

Dial the WorldPhone toll-free access number of the country you are calling from. Dial or give the English-speaking WorldPhone Operator your MCI Card number. Dial or give the WorldPhone Operator the area code and telephone in the US that you are calling. Or if you are calling from one country to another, dial the Country Code, City Code and number you are calling. If you don't know the codes, the WorldPhone Operator will assist you. In some locations, public phones require coins or phone cards to get a dial tone, Phone cards can usually be purchased at post offices and at many hotels and newsstands. Use a public phone if your hotel blocks WorldPhone calls.

Source: MCI Communication Corporation, 1995.

Since deregulation of the telecommunications market in the U.S., the PPCC has become the fastest-growing business. New companies and retailers have been engaged in direct sales—hawking, selling telephone, beepers, and other portable products in the streets. While retail shop owners and digital equipment stores may find that approach disadvantageous to their business, they may be benefiting from the hawking for the following reasons: (1) hawkers serve as advertisers because prospects see the products first hand, before deciding on their purchase; (2). Prospects tend to be skeptical of hawkers. They feel hawker's products are either too generic (non-original) or are illegitimate (stolen). Hence, purchasing a product directly from a store guarantees purchaser's security because it is accompanied by a receipt. Street-selling of prepaid telephone cards has, however, turned it into a multi-million dollar business. Most retailers have competitive rates for cards that can be purchased in convenient stores, fuel stations, banks, and post offices. The U.S. Postal Agency promotes a prepaid calling card with services to mostly European countries. Other retailers include PTI Phonetime and Econo and are available in most cities in the U.S. They have been able to attract customers by advertising affordable fares on posters, the calling cards themselves, and in the electronic media and by providing easy to follow instructions on the each card. Conveniently sized like a driver's license, the prepaid card includes instructions in English and Spanish languages, customer service toll free numbers for query resolution, and an expiration date. The calling process involves dialing a toll-free number, customer's pin number and then the number caller wishes to reach beginning with an area code. The caller is told by an electronically activated voice mail the remaining cash amount and time on the card prior to call completion. This entire process:

1. Helps caller in determining the length of time he/she wished to spend on the phone.
2. Prevents caller from worrying about expecting a phone bill or budgeting for a non-predetermined charge on the bill.
3. Helps caller to save money. Whereas a subscriber must budget at least $20 per month for regular service, excluding long-distance charges, a prepaid phone service requires a

direct one time payment. A consumer could have used a resident service only for several long-distance calls, however, he/she had to pay the an additional basic utility charge to maintain his/her service. Prepaid calling cards are designed only for long-distance(LD) calls and they cost $5, $10, $25, or $50 with the least amount valued at 15 to 25 minutes worth of calls.

4. PPCCs could be preserved as relics or souvenirs and could be auctioned at a higher percentage rate whereas phone bills could only be used as reference or research points.

The disadvantages of using PPCCs could be overwhelming:

1. Whereas the lowest charge for a call made from a personal telephone line was about 7 cents a minute depending on the carrier, calls made on prepaid cards cost between 25 and 40 cents per minute.
2. A call completion was worth 2 minutes, even if dialogue did not last that long-, however, PTI customers were charged for 2 minutes, upon call completion. As soon as a decoder answers the call, caller is charged 2 minutes, even if the decoder hangs up. Moreover, caller's card is invalid if it has 40cents or less remaining. Hence, a 25-minute card actually lasts 19 minutes or less.
3. Only callers with access to a touch tone phone could carry out the transaction. This policy discriminated against rotary phone owners. Moreover, a caller could not make calls whenever the need arose.
4. Overall, PPCC prospects were more likely to spend more money than regular business and resident customers. A resident customer on the MCI's Friends and Family discount program could make numerous calls for 15 cents per minute or 74 cents per minute when calling Cameroon. It cost approximately $1.20 per minute on any PPCC.

In conclusion, the PPCC market could flourish in developing countries because most people are not used to paying bills, and even when they do they do not pay it on time. Those who use facilities and services of companies are usually undependable. However, companies expecting to set up the business in Africa should negotiate with existing providers, especially the prospective country's former colonizer, who monopolize the market. Alcatel, the French telecommunication company has subsidized public telephone booths in Cameroon to local retailers who charge from 25 cents (100 francs, CFA), for intercity calls—calls made within a 300-mile radius—to $6 per minute (3,000 francs, CFA) for calls made to U.S. cities. U.S. companies can provide a more affordable bargain rate to that market if the business is demonopolized or opened to foreign markets. If a foreign company wants to capture a huge market in Sub-Sahara Africa, it should:

1. Train local residents, who understand their people's mindset, for technical and mid-level managerial positions.
2. Understand the people's culture through observation, in order to design effective campaign strategies. A persuasive ad should contain or imply knowledge of local humor, local character.
3. Use local leadership—those the people are loyal to—in the ads.
4. Use local music and popular events like soccer, to promote products.
5. Establish offices in major cities for customer services; inquiries and product delivery.
6. Set up or install cable systems and fax machines

U.S.-based Networks in Africa

AT&T's Africa One Program

As stated earlier in this chapter, AT&T has been installing a fiber optic network that will facilitate electronic communication in West African coastal countries by 2000. Part of its plan is to sell services to Post and Telecommunication (PTT) ministries. According to the deal, PTT will:

1. Buy capacity from AT&T, and train local workers;

2. Government will promote AT&T services through media campaigns, especially newspaper, radio and television advertisement. This effort will increase business potential in the suburban areas. A resident will be able to call the government office. Locally manufactured products like handicrafts and clothes will be shipped directly to buyers through telephone and computer transaction. Foreign companies and businessmen in Africa will be able to reach their families and affiliates anywhere. AT&T's fiber optic loop would carry fax and telephone calls, video and data transmissions to and from countries that are hard to reach. Also Afronet, like Africa One, is conducting a similar project in Africa which is already linked to Internet;
3. ATT will Expand existing businesses, nurture African markets, foster regional trade, and establish new trade partners.

Africa One has designated responsibilities for regional telecommunication authorities, to ensure that it maximizes profits and provide effective services. Regional telecommunication companies like the Pan African Telecommunications (PANAFTEL), RASCOM, and PATU, should be able to:

Operate and maintain their own network;

Levy cost-based tariffs to telecommunication subscribers;

Recommend cost-based end-user tariffs to company executives;

Distribute profits to investors; and

Retain some profit for equipment maintenance.

The responsibilities of telecommunication authorities are to invest or purchase capacity, work with partners to predict capacity needs, set prices and sell services to end users, and share rewards and responsibilities of ownership. But AT&T's responsibilities are limited to supplying the undersea network and integrating services with PANAFTEL, RASCOM, PATU, and national networks like INTELCAM and ZAMTEL. Also, transferring expertise in network operations, maintenance, billing, marketing, and owning equity in the network are other responsibilities.

AT&T anticipates huge benefits if others invest in its Africa One program. It hopes to see a rapid growth in tariff revenues resulting from increasing demands in telephone, fax, and Internet use. AT &T also anticipates public investment in order to increase its network services. The architecture of the program will be tri-tiered; the global tier which connects major geographic regions, the regional tier which connects countries within a geographic region, and the domestic tier that will connect population centers within a country. This global network plan looks practical, but problems might arise during its implementation. For example, not everyone will afford to have the equipment, and there is no guarantee that this equipment will be reliable, nor will there be prompt repair of equipment, given the history of negligence of telecommunication equipment.

Afronetwork, Inc.

This corporation is based in Delaware and its purpose is to assist African countries in modernizing their telecommunication services. In addition to Africa One service offers, Afro Network seeks to increase the number of telephones, mobile telephones, telex, telefax, teleconferences, electronic banking operations, and digital installations including fiber optics and laser communications. The Afro network project also includes the addition of four TV stations with 50 to 100 per country, and the construction of a quality technology and hence ameliorate the balance of payment problem which Africa has experienced since its colonization.

Summary

Not only would telecommunication programs facilitate business transactions between countries with small economies and those with powerful economies, but they would also bridge the cultural gap between those countries. Although, there is reason to suspect that through this global practice, major corporations operating in different regions would become the real ruling force because they have the ability to influence local politics and economy by giving jobs to residents. A sound economy reflects a stable government. In fact, morals, good health, and wealth are perpetuated by a balanced economy and there are less political crises in any society where wealth and money are readily available to the people. These people are less likely to fight for political changes if they have an equal chance to have money and wealth. That philosophy of life which started the Industrial Revolution applies to all countries, including those with communist regimes. The fact that China seeks to strengthen trade relations with the "free" world only shows why the Soviet Union abandoned communism for a market economy, and why smaller states like Cameroon and Haiti struggle to replace communal liberalism with the sort of democracy that promotes self empowerment Although this new way of living is costing developing nations like Nigeria, Kenya, Haiti, Bosnia, and Lithuania human lives, many countries will benefit from a free market economy. Does that mean America and other market-oriented forces should indiscriminately transfer large volumes of communication technology or selected equipment to developing countries? If so so, who will benefit the most from this practice, the exporter, importer, or user of this equipment? In the chapter ahead, such issues are discussed.

Chapter Three

African Communities: A Case Study

This chapter examines the information market in African communities in Africa and in the United States.It further samples the ethical, moral, and market capabilities of information consumption and describes the psychological contexts for determining larger markets and provides a background on the transfer and use of communication technology by Africans abroad and the degree of interest in using the technology by target groups in Africa. This sample frame was selected to represent the views of African telecommunication prospects and users based on the following assumptions:

1. Business transactions become successful only through the input of the prospect and qualifier, as well as the provider. To determine the extent to which their products can be marketed, providers need information regarding the prospect's needs and tastes.
2. Certain subjects targeted for this inquiry have some experiences using technology. African diplomats living in the U.S. have been exposed to the technology more than any other African communities because the U.S. has the most advanced telecommunication system in the world.

3. African-based subjects will show an enthusiasm for providing information about the importation of telecommunication products because they have demonstrated such an interest in other foreign products.

Background of the Study

The Political Image of Africa

Many foreigners see Africa as a country, not a continent. Only a minority of Africans—the educated, who make up about 24 percent of the entire population—are aware of its political state. This elite group nurtures only a basic knowledge of geo-cultural and economic parameters in most countries, gathered mainly from classroom lectures, local newspapers and newsmagazines. The uneducated are not familiar with any community beyond the sphere of their landscape and culture. Similarly, people of other races, including blacks in the Diaspora, have formed negative concepts of Africa; that place of famine, war,poverty, and unstable governments. Knowledge of Africa is based on insufficient information provided by the media, and ignorance, indifference, and denial of Africa's potentials prevail in the international community. Africa's political history is characterized abroad by slave trade between African kings and merchants from the Northern Hemisphere, and its colonization by Europeans has left an indelible impression in people's minds about the continent's sociopolitical state. Arguably, these notions have prevented foreign governments and business organizations from considering African leaders and business elite as serious parties in global decision-making. Consequently, they have been regarded as posing no threat to the course of international polity, nor as viable contributors to world science and technology. Although some of them hold significant power in world organizations, African leaders are seen as followers. They have not had any significant input in shaping the course of world events. Despite their nomination as UN Secretaries General, Boutrous-Boutrous Ghali and Kofi Anan have not been able to galvanize world leaders to successfully resolve Africa's crises. During his tenure, Ghali could not convince the UN to resolve political crises in Somalia, Burundi, and Liberia, nor did Anan advocate policies that would curb the castigation of Muslims who have outnumbered other religious groups in Africa, the Middle East and Bosnia.

Western-backed African presidents have been treated with a similar cynicism. These governments provide African countries with financial, military, technical, and diplomatic assistance, and resolve social, economic and political crises caused by war. Former U.S. Secretary of State, Warren Christopher once declared that America would only help a country when it has a vested interest (*Time*, 1994). This capitalist concept of help does not surpass the machiavellic perspective of Africa generated by foreign scholars who have distorted Africa's history, for ulterior motives. The motives include miseducating the world about Africa's real past in order to promote distrust toward Africentric scholarship and undermine black achievements. Only Basil Davidson, Boubacar Barry, Kiz Zerbo, Kenneth Dika and Ivan Sertima rank among legitimate historians who have captured facts about Africa's past. Black anthropo-historians like Maulana Karenga, Cheikh Anta Diop, Molefi Asante, and sociologists like Harim Jean, Pierre N'Diaye, Engelbert Mveng, and Pathe Diagne have provided significant information about African societies.

The writings of the aforementioned have successfully created a context for defining Africa as a potential marketplace for ideological, infrastructural, financial and economic investment. Nwabuzor's (1980) study (there has been no other significant one) showed an ability for the Tikars—a numerically dominant tribe in West Africa—to sacrifice their values for foreign ones. The Tikars and the Masai can think independently with or without such thought-inducing products as radio and television programs and videocassettes, because of they adhere to their cultural values. However, in Ibo, Akan, and other societies that had European missionaries and administrators for decades, independent thinking has been impossible, for reasons discussed later in this book.

Africans influenced by foreign culture have been selected for this study. They were classified at two levels:

1. The public: business men and women, and educated ones. The term *educated* was a diacritic for anyone who could speak and write a European language. The underlying concept here was, a user of any European language had the ability to form opinions about European things, telecommunication products included.

2. Embassy staff: representatives of African governments serving in their respective embassies in Washington, D.C., was selected instead of African students and professionals in the U.S. because it represented the policy making institution in Africa. Hence, it was perceived to have informed opinions regarding the transportation to, and use of telecommunication, in Africa.

Respondent Traits

As with survey methodology, data gathering for this study was a challenge. Respondents in African embassies proved more difficult to release answers than those in Africa. They were impatient to answer questions regarding their views on transferring telecommunication technology to Africa. Some embassy staff were reluctant to provide basic information, including addresses and telephone numbers of telecommunication ministries in their countries. African-based respondents were more eager to answer questions. Some of them wanted to know whether such equipment would be available to be purchased.

Questions for Embassy Subjects (Instrument 1)

The questions in this schedule sought to determine:

1. Their use of telecommunication equipment in the U.S. The questions sought answers to the quantity of equipment they were using, the amount of time they had been using it, and whether they would send the equipment to Africa. This set of questions intended to check their knowledge of, and experience with, the products. They also sought to determine the importance prospects attached to the equipment and their potential in serving the same purposes when transferred to Africa. Although the African market was not expected to react in the same manner as Africans abroad, this study assumed that reactions would be the same, given the same conditions of equipment availability and time length of their use. The only apparent differences between Africans at home and those abroad are exposure to and purchasing power of equipment.
2. Other questions sought to determine the subjects' views of their country's ability to manage the importation and operation of the technology. If an embassy staff used the equipment for infor-

mation transmission, that entity could advise others as to its importance. Cognizant that a political system cannot be stable without a stable economy, the study sought to know the extent to which government representatives would be concerned if major international telecommunication corporations brought their products and services to Africa. Would African governments lose their autonomy to the corporations if such corporations engaged in the massive employment and maximization of the economy? The rationale for raising the question was twofold:

(a) To determine whether embassy staff were still loyal to their government. One can assume that government officials abroad who do not care about their country's sovereignty would support the uncontrolled flow of foreign information technologies and services. On the other hand, loyalty might be stifling; a loyal or patriotic person might despise agents that bring progressive change. Along the same theoretical praxis, the role of a government representative is to promote the government's ideals, and since most governments advocate (in principle) ideas and practices that cause progressive change, they are expected to support such ideas and practices. If a process leads to economic improvement and better living standards, concerned government officials should express their views on the issue.

(b) The question considered critical to the study, asked government representatives to rank their responses regarding their country's participation in the global information supermarket. It sought to know whether they would like their citizens to pay for access to electronically processed entertainment and economic information.

Questions for the Public (Instrument 2)

Instrument 2 sought answers to the commercial need for products. In constructing the instrument, the study assumed that the African public (educated and business people) would be interested in using more advanced technology to enhance their interactions. Public was defined as any educated or business person. Educated ones were those who could read, write and think in a European language. Following the installation of a fiber optic network in West African coastal countries, it was thought that business men and women would have a need to express their views on the socioeconomic importance of the project on their living standards. Hence, most of the questions designed for the public sought to know the best means through which they reached their business partners locally and internationally, the kind of communication equipment they used, and the quality of services. One of the questions sought answers to the number of attempts made by a customer to complete a fax or telephone dial. It did not seek reasons for their ranked choices because such would imply prying into the residents' privacy. Although previous studies on African communities have indicated responses to "personal questions" (Nwabuzor, 1980), respondents have deliberately avoided questions which check consumption values (Ngwainmbi, 1991). Moreover, Africans have a culture of suspicion. Their leaders virtually dictate to the people. This dictatorship has suppressed the ability to express themselves freely, especially on matters related to the entire community. Village residents are not allowed to provide strangers with information; e.g. addresses or directions. They appear shy and only direct strangers to a dignitary who has the vested authority to handle such requests. Therefore, questions asked about how much they knew, or the means by which they acquire knowledge may be construed as a conspiracy against the government. This is why most Africans do not want to participate in unsolicited interviews or questions related to government matters. A pilot study conducted between 1987 and 1990 which sought to evaluate government performance indicated limited response (Ngwainmbi, 1991). Out of about 140,000 government workers contacted in their offices in Cameroon for a survey, only 26 participated. Similarly, among the over 150 embassy African staff in Washington, D.C., only 26 participated in the interview, despite at least three previous trips made to each embassy. On the other hand, the public showed much enthusiasm toward participating in the survey. The basic rationale for this response rate disparity is, the African public has, for decades been exposed to foreign-oriented freedom ideas and they are under no specific constraints to concealing individual views on general public matters, especially if such views can help in the improvement of their economic and social living standards.

Following their countries' independence from colonial rule in the 1960s, most African governments have not been able to control an infiltration of European values. These governments had relaxed policies on the importation of art, print and broadcast media, and other social habits. In addition, African parents have been advocating Western values by sponsoring their children in European schools. As such, Western education has become the yardstick for measuring economic and social standards. This new phenomenon has created a hybrid culture—an elite class. Important people in most African societies are not kings or traditional rulers, but the educated and the wealthy. Educated people constitute the growing majority of the African population, partially because government, private, and missionary institutions provide low cost education in remote parts of the continent, and there is no age limit required to receive education. Since 98 percent of the African public approached participated in the survey for this book, suggests a high level of enthusiasm toward the acquisition of foreign knowledge. Respondent ages ranged from 16 to 64. This sample frame was randomly selected to determine the elasticity of the telecommunication market. The younger respondents were expected to show more enthusiasm in advocating the importation and use of the products, rather than the older respondents, as the former interact more frequently with foreigners. The questions were designed to seek specific views on the importation and use of telecommunications.

Study Limitations

The results of this study should not be used in evaluating, or making decisions about the African telecommunication market for the following reasons:

1. The sample frame is small, therefore, it is not totally representative of the entire population's views. Due to time constraints, transportation difficulties, and language translation difficulties, only 152 people participated in the study. Hence the internal validity of the study is inconclusive.
2. One day was spent in each country to interview subjects, hence more subjects could not be reached. The probability of interviewing respondents two or more times was high, especially since the study did not track respondent demographic information. Additionally, because they were interviewed in non-specific areas; homes, business sites, and streets, extraneous variables may have infiltrated the study, that is, respondents from different streets or countries may have moved back to the

previous interview site. Nevertheless, because the questions were designed using Likert Scale to seek specific answers, results may be used in determining a new market for telecommunication products and service; and

3. Only respondents in ten of the fifty African embassies participated in the study.

Survey Results

As a reminder, the questions designed for the public sought to determine their level of interest in importing and using communication technology. Given that African residents rank among the poorest in the world, it would be unlikely to find many people with a vast knowledge of telephones, fax machines, beepers and satellite dishes. Hence, the study posited that most telecommunication users would be business men and women. The first question differentiated international business respondents from non business respondents. The assumption was, only international business entities were wealthy enough to own and use telecommunication equipment in doing business internationally, exclusively. Survey results showed that 61.9 percent did business abroad, while 38.1 percent did not. The study proceeded to know how they reached their partners abroad. This question was designed specifically to determine the rate of feedback and as a latent cause, the need for the importation of additional products. Six variables described means of communication; fax, cordless phone, telephone, letter writing, making trips abroad, and other means. Eighty-six point five percent said they reached their foreign partners by telephone, while only 0.8 made trips abroad. A considerable 2.4 percent used cordless phones and only 13.5 wrote letters (see Table 3.0). The study further sought to compare the percentage of international business makers with the modus operandi. Of the 86.5 percent who did international business, 57.2 used fax, as opposed to 11.2 percent who wrote letters. This shows (1) the interest in communicating by the fastest means, and suggests (2) the increasing demand for telephones and cordless phones. Although only 25.4 percent used facsimile, the product is in higher demand than letter writing and trip making. Although the study did not ask respondents about the rate of writing business letters and/or of making business trips abroad, such means may have been less reliable and less profitable for the African business entities, as only 4 percent admitted to using both approaches.

Telecommunication products are also useful for local commercial and non-commercial transactions. Not all business people have to do international business in order to generate a need for using telecommunication products. The study sought to know how local-oriented business people reached their partners. Among the 38.1 percent who were not involved with international business, only 0.8 percent said they contacted their partners by sending messengers. The study went on to find out how long it took to complete a fax or telephone call. 77.8 percent said it took one attempt to complete an international fax transaction, while 43.7 percent said it took more than two attempts. Given that telephone and fax users use the same lines, it seems startling that fax dials went through faster than telephone calls. However, since there are few international telephone lines and fewer fax machines, one would expect telephone callers to experience more difficulties in completing their calls. The lines become congested as more callers clamor for few lines. Conversely, because fewer fax machines are available only to fax owners, there is a low probability of experiencing dialing congestion. Moreover, African-based fax users may not be dialing to the same high density zone as African-based telephone callers whose primary destination is the US. Additionally, fax owners may be using digital encryptions.

The study also sought to determine the types of products imported by business men and women. The purpose of this inquiry was to know the product(s) in high demand. Surprisingly, only 11.9 percent said they imported telephone sets, as opposed to 34.9 percent who imported radio sets. 3.2 percent imported computers, 38.9 percent television sets, 2.4 percent fax machines, and 19.8 percent satellite dishes. Only 0.8 percent imported telephones sets (see Table 3.1). Certainly, the demand for beepers and fax machines was low; this may be due to the limited number of free telephones. Since all calls are charged, importing and/or using beepers or fax machines would require the installation of more affordable telephone lines. The high demand for television and radio sets and the limited demand for beepers and satellite dishes show a greater consumer interest in obtaining news and entertainment information than in using mobile telecommunication products for business purposes. The fact that 90 percent and 85 percent of the business respondents ranked radio and television sets respectively as their customers' most preferred

Table 3.0
Frequency layout on communication channels by international business respondents

Value label	Percent	Cum percent
By telephone	86.5	90.5
By fax	2.4	92.9
By letters	3.2	96.0
By making trips	0.8	94.4
By cordless phone	2.4	80.2
By other means	3.2	99.2

product enhances the concept that the broader spectrum of the African telecommunication market has not discovered the commercial importance of telecommunication products. Consumers are still interested in using telecommunication systems as channels of social rather than business communication. They do not use these systems to generate profitable business—money. Hence, they can be considered a consumer- rather than profit-oriented market.

Through a crosstabulation, the study sought to determine the level of their commitment in transacting business with foreign partners. The assumption was, any committed business entity involved with a foreign partner would need or have more efficient communication tools. The study showed that only .021 percent imported fax machines, telephone sets and computers, respectively, as opposed to 30.9, 32.9 and 13.8 percent who imported satellite dishes, television and radio sets, respectively. Since there was such limited interest among international business enti-

Table 3.1: *Equipment importation by businesses*

Value label (imported product)	Percent	Cum percent
satellite	19.8	41.3
television sets	38.9	80.2
radio	15.9	96.8
fax machines	2.4	94.4
computers	3.2	97.6
telephone sets	0.8	96.8

ties in importing telecommunication products, did local businesses and educated ones import any telecommunication products? Was there a shift in the demand for foreign products? Among the 38.1 percent (48 respondents) who had no business dealings abroad, 28.9 percent imported radio sets, 14.5 television sets, and 9.9 percent telephone sets. Also, 2.0 percent and 2.6 percent imported fax machines and computers, respectively. Since more locally-oriented business and educated people imported more telephone sets than international business people, one may conclude that the high demand for telephone sets may have been caused by the increasing need for socializing by phone, as earlier propounded—a demand not generated by business ideas. Moreover, as only 1.3 percent of those doing business abroad used beepers, it is clear that the desire to carry out fast business transactions with international partners is limited. Perhaps, the dramatically low importation of beepers also suggests either limited knowledge of its existence, or its importance in enhancing communication.

The African public is enthused about information technology, but it has not been adequately exposed to the importance of the technology in enhancing economic activities. Rather, the larger market has demonstrated an interest in using communication technology mainly for entertainment. Hence, product providers and communication practitioners should educate the public on the cash value of information technology.

With regard to Instrument I designed for embassy respondents, the study assumed that diplomats living in the United States would be more knowledgeable on the use of IT in mediating commercial and entertainment messages, because they were more exposed to the technology than the African public. Hence, questions dealing with the quantity of equipment owned, time of ownership, and whether they would ship the equipment to Africa did not only seek to determine the importance of the technology to their daily lives, but they also sought to determine its socioeconomic import to Africa's development from the government's angle.

The first question which requested respondent profession was aimed at identifying profession with attitudes toward IT use and transfer. The assumption was, diplomats trained in academic fields like psychology, education, and economics would be more accurate in presenting views on Africa's participation in the global information economy, than politically inclined diplomats like cultural attaches, military officers, ambassadors, and counselors. Additionally, the higher the number of telecommunication products owned and the greater the number of years spent in using it, the greater the knowledge of its importance to Africa.

Nine categories of respondents; teachers, accountants, diplomats, secretaries, cultural attaches, counselors, business men and women and telecommunication technicians, answered questions ranging from ownership and purpose of use to concepts on effects of importing and using telecommunication products and services. The purpose of identifying product ownership was to help sales agencies, brokers and regulators determine what to sell. Ownership frequency chart revealed that 92.3 percent owned television sets, 80.8 percent VCR, 42.3 percent cordless phones, 80.8 percent telephones, 23.1 percent fax machines, and only 3.8 percent beepers. Among them, only 3.8 percent of the respondents used the products for business with a 42.3 percent entertainment consumer market. 26.9 percent used it in transmitting information and only 7.7 percent found the products useful for both entertainment and information transmission. Since diplomats are primarily involved in intergovernmental and inter-corporation business, a majority response was expected on information transmission habits. Only 3.8 percent used personal IT products in transacting business. They were more interested in sharing the product, as 76.9 percent of those who owned fax machines, television, VCR, telephone, computer, satellite dishes, and beepers said they would ship them to Africa. That strong response also suggests governments interest in supporting the circulation and use of electronic information technology and services. All respondents wanted their country to become a part of the global information super market. The study instrument had described "information supermarket 'as having access to business, social, and economic information from a database. To validate their views, the extent of their loyalty to the government, their patriotism and clairvoyance to global reality, the study requested answers to their degree of concerned about their country's ultimate loss of autonomy to more industrialized nations. That issue was an integral part of the study because most educated Africans have been skeptical about Africa's dependence on foreign assistance, especially regarding the importation of technology and expatriates. They have, indeed, considered dependence a deterrent to the maximization of local incentive.

The responses were ranked in degrees of concern. Responses showed that 69.2 percent of African educated class still nurtured the skepticism, while 11.5 percent were not concerned. This minority percentage which was oblivious to the issue of Africa's autonomy represents the size of enlightened Africans who foresee benefits in using new information technology in African communities. To verify the validity of this fact, the study compared embassy staff professionals with views on the exportation of IT to Africa. A crosstabulation comprising nine different types of embassy professionals and their views sought to determine whether the better educated staff; teachers, accountants, PR Officers, Counselors and Diplomats would support the exportation of IT products, than the support staff; secretaries, military officers, receptionists, and cashiers. Better education was equitable to having better ideas and better decision-making potential, while "support staff' was considered less likely to operate as well. Similarly, for the purpose of this study, diplomats, counselors, financial experts, and PR officers were deemed to have the most influential positions in the embassy because they made official decisions. Hence, their responses as to the impact of exporting and using IT in Africa ought to be considered seriously by foreign providers, governments, and consumers. Surprisingly, receptionists ranked highest (52.6 percent) followed by diplomats with 36.2 percent . Counselors who par excellence, are the conscience of ambassadors and who should demonstrate a strong knowledge of new world trends (including the social and economic advantages IT to national development), appeared far less enthused about encouraging the shipment of information products to Africa, as only 3.9 percent supported the exportation of IT (see Chart 3.0).

The theory that counselors, a part of the elite class in Africa, are more knowledgeable about their country's development needs has been challenged. Conversely, this study has shown that Africans are more minded toward goods and activities that support progressive change as 86.3 percent advocated the importation of IT products, despite having been less exposed to it than embassy counselors. Even diplomats and PR officers whose duty is to mediate between international governmental and business communities and their countries, in order to foster international understanding and "market" their country in a profitable manner failed to foresee the benefits of importing IT.

But what were the reasons for discouraging IT shipment? 40.8 percent said they were very concerned their country would lose its autonomy to industrialized countries, while only 10.5 percent did not find autonomy forfeitment a problem. The study further asked respondents if they preferred an Africa-based and an African-owned industry to

Chart 3.0: Support for IT Transfer to Africa

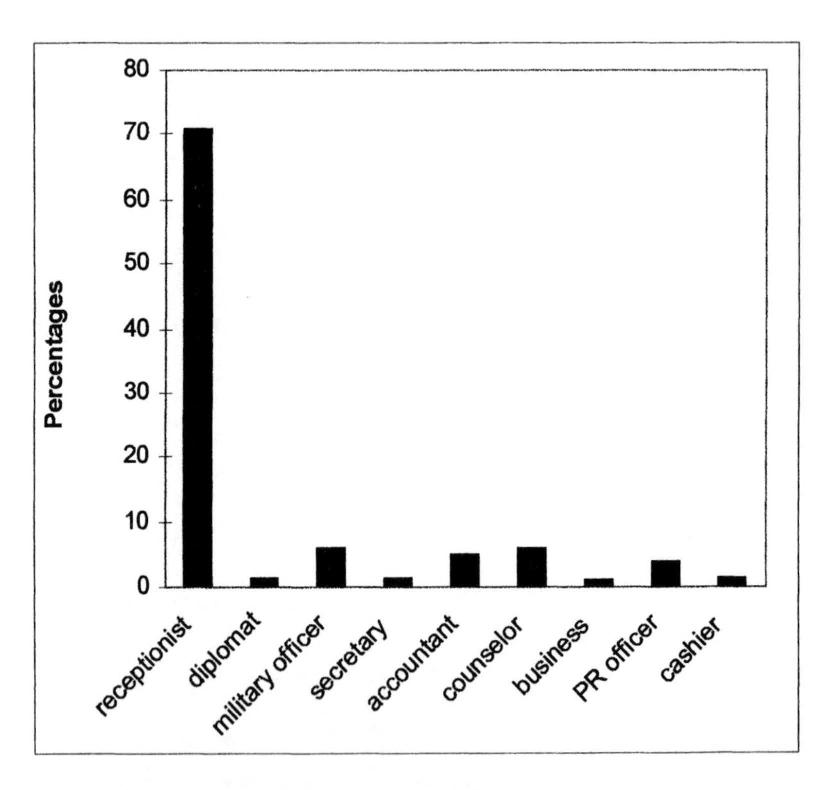

and fax machines. Asked why they preferred internal production to importation, that percentage of respondents responded as follows:

1. To promote African-based industries.
2. To reduce the outflow of currency.
3. Proximity and provision of training of Africans.
4. The process will be cheaper and more accessible to most Africans.

4. The process will be cheaper and more accessible to most Africans.
5. The process will increase employment and economic security and it will ensure sovereignty.
6. The process will encourage reliance and interest in the use of local products.
7. It will reduce cost, and
8. It will improve revenues.

Even among the 3.8 percent who had totally rejected the idea of shipping IT to Africa, diplomats, and military officers wanted Africa to manufacture its own products in order to curb the depletion of funds, and to determine the pace of development. Military officers discouraged the transfer of IT because of a voltage difference in gadgets suitable for Africa. The most active advocates for internal production were receptionists (70.4 percent) and the least interested were secretaries (0.7 percent) as shown in Chart 3.1.

The study attempted to find a correlation between patriotism and time spent in the United States. This search was aimed at determining whether:

1. Time spent in the United States affected the diplomats' views of their country's political abilities; and
2. Ideas of democracy, a free market economy, and free speech germane to America and reflected in the massive availability and widespread use of electronic information systems, the increase in IT markets and providers, and the free circulation of electronically processed information influenced their views of their country's economic and political climate.

The search was also aimed at evaluating the extent to which time spent in the U.S. impacted on their knowledge of civil society (a place with a potential to improve). Results showed that those who had only been in the U.S. 6 months or less expressed the greatest concern. Only 4.6 percent of those who had lived in the U.S. for more than six years were very concerned. This shows that the greater the exposure to telecommunication products, the more free minded the individual. Having identified the idealistic responses of African government officials serving abroad, the study proceeded to analyze respondents' views on the eco-

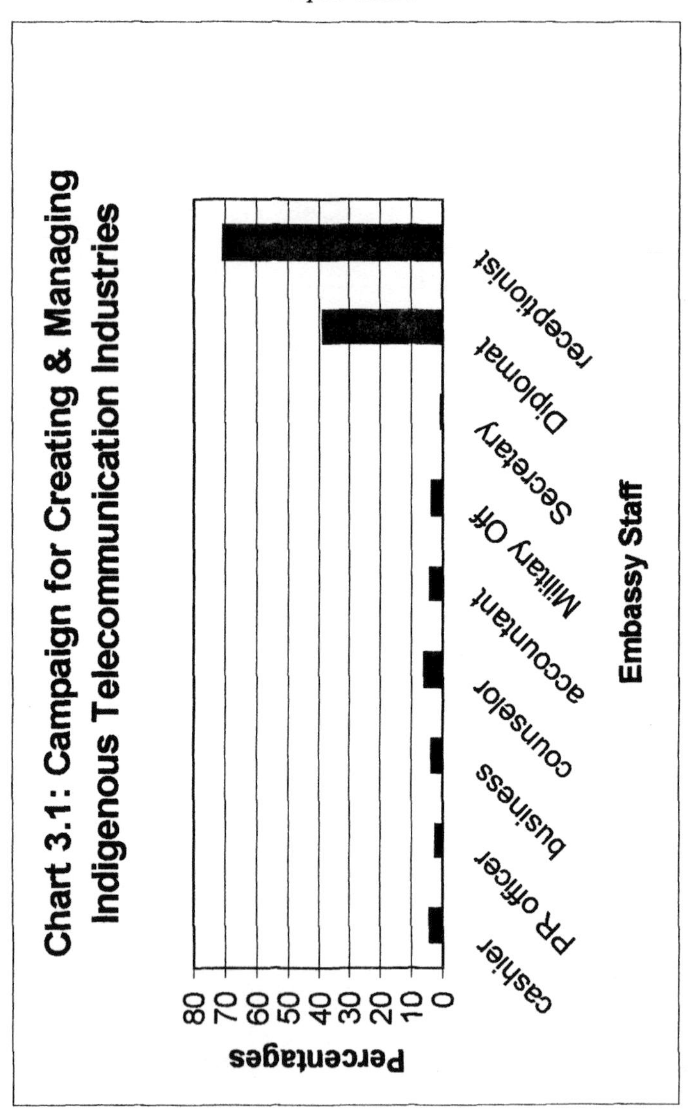

juxtaposed with product ownership to help providers in knowing which products would be more marketable. Analyses focused on support staff who admittedly owned more telecommunication products than the better educated staff. 52.6 percent (receptionists) owned television sets, 36.2 percent telephone sets, 2.0 percent computers, 3.9 percent fax machines, and 1.3 percent satellite dishes. The study then cross-checked amount of time staff owned a product with their views on their country's

chines, and 1.3 percent satellite dishes. The study then cross-checked amount of time staff owned a product with their views on their country's participation in the global information revolution. Only those who owned more products would be enlightened enough to discuss its social and economic impact. Results showed that 70.4 percent of those who had used the equipment between two and five years mostly wanted their country to become a part of the global information market. Ironically, only 5.9 percent of those who had used for more than six years strongly supported that view.

The study tested respondent length of stay with opinions on country participation to accentuate the microeconomics theory of investment. A longer length of stay presupposed greater knowledge of investment in information technology (IT). Similarly, since access to IT is relative to experience, embassy staff were deemed more capable of providing informed opinions on its usefulness in the transaction of business messages than the public. Results revealed that 70 percent of those who had only been in the U.S. for less than 6 months strongly supported the idea of investing in IT, while only 5.9 percent of those who had lived in the U.S. between 2 and 5 years softly agreed. The differences in time spent abroad affected views on national participation. The study compared the length of time staff had owned the product with their views about their country's participation in the New World Information Order (NWIO). (In the 1970s the United Nations set up the MacBride Commission to oversee the worldwide circulation of media information. One of the concerns raised by Third World communication experts is the lack of capacity including technology in Third World nations, to compete with industrialized nations on the collection, processing, and distribution of electronic information.) The reasons for this inquiry were, (a) only subjects with the most experience on IT use would fully express a better understanding of the benefits their country would have if it invested heavily in information technology, and (b) A country's full participation in the Order could increase its economic benefits and enlighten or broaden its citizens' outlook on world issues. Again here, government officials, especially embassy counselors and diplomats (who have policy-making potential) were expected to respond positively. As long-term users of information products, and as international negotiators for their country they were expected to be the strongest advocates. Moreover, they lived in the US—an information society with strong economic capabilities. Hence, because of their experience with the use-

showed that 70.4 percent of embassy staff who had used the products between 2-5 years ranked the highest in supporting a national IT campaign. Of those respondents, 78.9 percent were receptionists, as op-

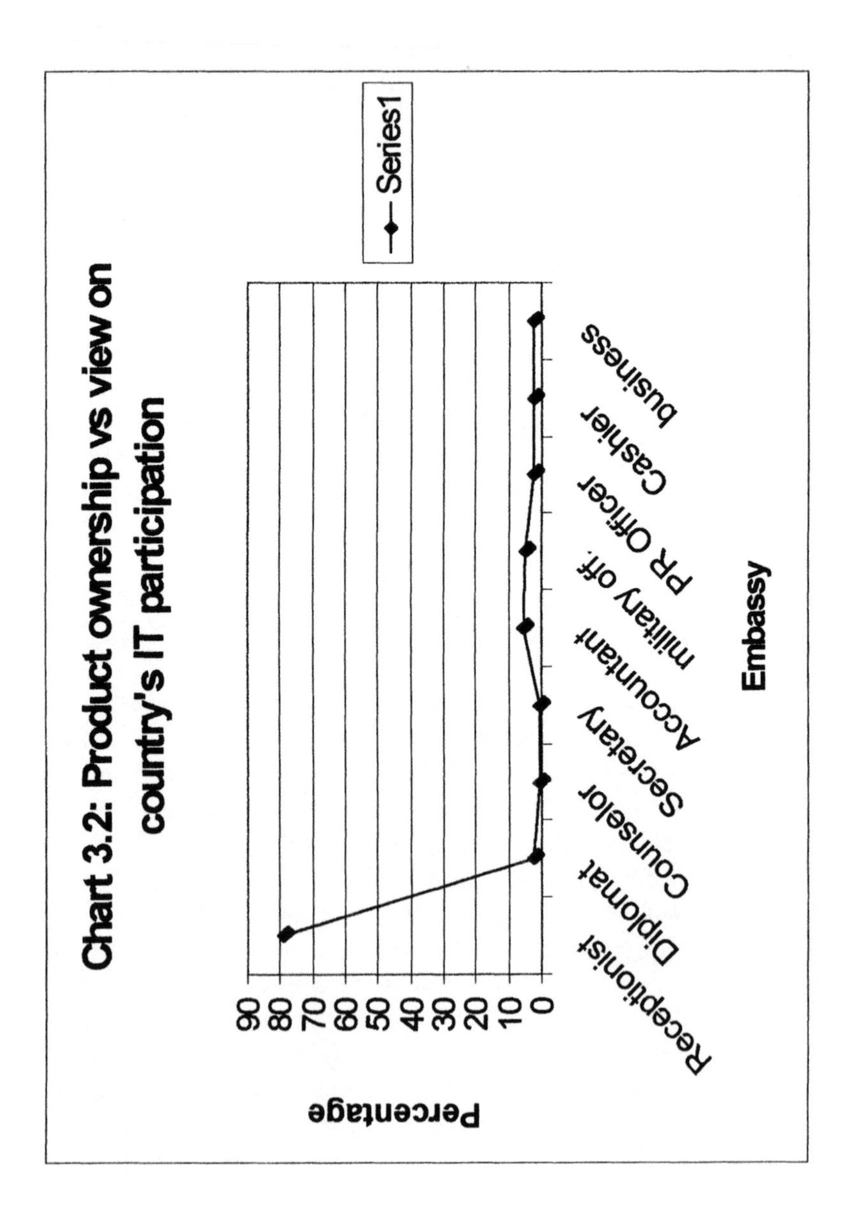

Chart 3.2: Product ownership vs view on country's IT participation

paign. Of those respondents, 78.9 percent were receptionists, as opposed to 2.5 percent diplomats and 0.6 percent counselors (see Chart 3.2).

Clearly, policy-makers were far less enthused about recognizing the importance of using information products in the promotion of economic development and social mediation.

Summary

Embassy respondents had diverse and skeptical views on the socioeconomic and political impact of importing and using IT in African countries, while the Africa-based public unanimously embraced the idea. As expected with diplomats, responses differed on direct questions. Nevertheless, their support for the exportation of IT to Africa shows a changing mentality toward foreigners and imperialism, despite the prevailing view of other African elite groups. Conversely, Africa-based respondents did not pose a surprise with most of their responses. Due to their lengthy experience with colonial rule and their exposure to, and experience with, with foreign technology and foreign customs, they were expected to, and did, encourage the exportation of IT.

Providers, policymakers, sales persons, and communication experts should reevaluate the needs of Africa's telecommunication market and serve it accordingly. Although inquiries into respondent financial capacity was visibly absent in this study, considerable description of Africa's purchasing power has been provided in latter sections of this book. Corporations and service providers are reminded this study is a representative sample of Africans' views; only Cameroon, Sudan, Algeria, Guinea,Tanzania, Zambia, Senegal, CAR, Chad, Benin, South Africa, Zimbabwe and Kenya embassies participated in this study. However, because the countries represent each region, corporations may find new markets for their products in Africa. Additionally, each country has its own telecommunication policies which include the importation and management of information technology and services (See chapter 6). Hence, corporations should understand a country's policies before engaging in brokerage and other business dealings with African prospects.

Chapter Four

Africa and the Information Supermarket

Overview

Perhaps the first attempt at creating an information supermarket in Africa was initiated in the 19th century by British voyager and philanthropist Cecil Rhodes. Rhodes, after whom Rhodesia (now Zambia and Zimbabwe,) were named, had established a railroad from Cairo to Capetown, to be followed by telecommunication lines. But that did not happen. Following the partition of Africa in 1884, which led to the establishment of new national borders, new governing techniques and infrastructures, the new governors (Europeans) sought to maintain communication with others in different African regions and at home. Telegraph wires became a viable means of communicating messages between law makers and their representatives in the colonies, and for mediating sensitive information in order to ensure the control of the region. During the Jameson Raid in the late 1900s in then Cape (now Capetown, South Africa), the British government in London sent a number of telegrams to Rhodes advising him of its position regarding the future of the Transvaal region. According to a telegram from H. Hanning on 2/8/1895, the British South Africa Company (BSA), discussed the financing of a 60-miles railway construction in the Transvaal region (*History Today,* June & July, 1962). The purpose was to provide cheap transportation for shopkeepers, agents, farmers, and pot merchants to Johannesburg markets. Railway lines also served as a new source of creating public intimacy, increasing business, and promoting cultural diversity. People with little knowledge about others in far away towns could go there, sell or buy products and services, share their values, and obtain services by using gestures and by interpreting their local languages. Some of the railway systems are still being used for the same purposes in Cameroon, Kenya, Nigeria, Tanzania, and other northern and southern regions. Their existence remind residents of the quality of European civilization. They provide affordable transportation to the poor. To some, however, they are remnants of colonial influence and imperialism.

The railway system was also an incentive for the installation of telephone lines in the administrative headquarters, primarily for regular consumers; colonial administrators, missionaries and their families, to communicate with other settlers—the British. Later on, the British built roads and bridges and eventually, radio stations were erected to spread governing and religious philosophies. This media campaign was effective in creating public awareness and in bringing European values to the literate African—a rapidly growing class throughout the continent.

Since the beginning of the 20th century, modern information technology has been used to spread messages beyond local boundaries and to enable the functioning of the global society. There are four main factors which determine the functioning of an information supermarket; (1) the supply, and (2) consumption of information, (3) its application modes and (4) the effects of its use. Described as a place where people seek, use, share, retrieve and store electronically mediated material, the information society has become the foremost market for the sale of ideas globally. Information has become a major commodity, used by corporations to promote their products and services and to generate profit; and by governments to influence public opinion and control international trends. A government with a strong satellite system can monitor developments in another country and use the information to its advantage. Similarly, the general public has access to information systems for carrying out entertainment activities which include generating dialogue, planning conferences, playing games, sending and obtaining current information, or transacting small-scale business. In industrialized countries, a household has an average of three information systems; the commonest of which are radio, television and telephone. Although they are used mostly for entertainment, in some cases owners have managed to generate profit. For example, public relations firms pay home-based media audiences to record specific characteristics in a number of programs. The firms then collect, analyze and sell the information to brokers and companies.

Given the multiple uses of information technology and the increasing demand for information, providers have been designing newer, more portable, and more affordable technology. As the world approaches the 21st century, satellite dishes, facsimile machines, telephone lines, modems, and the internet take messages to different people in different regions. Those systems provide easy access for information-sharing. Indeed, the information superhighway is basically a state where anyone can have access to information about anyone else. This suggests that for people in developed and underdeveloped societies, there will be virtually no secrets. The loss of privacy will pose serious adjustment problems for people in developing societies who operate in a secretive culture.

Becker (1994) explains this phenomenon properly in his paper: *Acculturation and Technology Transfer* by citing examples of the acculturation of European communications technology in non-European cultures. Contrary to Weber's (1976) philosophical assumption that rituals, religious and magical viewpoints play no role in the adoption of economic ethics or in the inculcation of foreign economic values when technology is transferred (p. 812), Becker maintains that developing countries do not only include the names of foreign technology in their linguistic repertoire but they also "adapt strange technical products to suit their local requirements" (p. 10). Given this practice, one would expect Africans to not only accept the new technology of the information supermarket, but to use it vigorously. Africans use names containing elements of foreign technology like "Lufthansa", "Coca-Cola", "engineer", and "telephone" which clearly indicates technological acculturation, reminiscent of the music, love-making and dress, capitalist attitudes, and socialization patterns that prevailed in the colonial era in Africa. Their contact with missionaries, voyagers, merchants and bureaucrats following Africa's partitioning by Western imperial powers, broadened their perspectives and made them more receptive to foreign goods than toward locally manufactured ones. Hence, their dramatic acceptance of things foreign—Christianity, clothes, transportation—will never change because things that come to Africa from foreign regions, no matter what their pitfalls are religiously accepted. When members of a culture quickly adopt foreign technology in order to recognize structures and use them as prescribed by their are religiously accepted. When members of a culture quickly adopt foreign technology in order to recognize structures and use them as prescribed by their inventors, they are said to be practising "rational capitalism". Whenever they become fascinated with foreign terminology, they subject themselves to the culture of the technology and by embracing foreign technology, people in developing countries tend to underlook the efficiency of locally developed technologies. African countries, which constitute the majority of the developing world, undermine their own locally-produced technology, and come to respect foreign technology for reasons provided in this and the next chapter. To further understand how Africans regard imported ideas and how they would deal with information technology (IT), a background of the fundamental cultural differences in African regions is necessary. Four paradigms of culture determine the potentiality of the African IT consumer: language, cosmology, ecology and colonialism.

Language

There are about 1000 language groups in Africa. These groups have Semitic characteristics, meaning they are linguistically related to Arabic and Hebrew. Arabic or Arabo Berber is used in Northern Africa while Negro African languages are used in sub-Saharan regions. Differences in civilization have influenced the languages. Due to the numerically dominant presence of Europeans in the sub-Saharan region compared to the Northern region in the colonial era, new diction and idiosyncrasies were incorporated into sub-Saharan languages and that produced a hybrid set of EuroAfrican languages, mainly *petit negre*, creole, and pidgin. Since the colonial era, educated Africans have never held lengthy conversations in their native languages without mixing in words from European languages. Africans in Eastern and Southern regions form the largest group of English users because Anglo-Saxon culture first spread from Egypt to Capetown.

Hybrids of French and English like *petit negre and pidgin,* German, Portuguese, and Spanish are the dominant colonial languages used by most educated people in the region, for official, instructional and discussion purposes. This group can only communicate most effectively through European languages.

Using these colonial languages is, for some of them, a mark of distinction—it brings respect to them. Since the psyche of the colonized African has been so infiltrated, it would be difficult to create an authentic Afrocentric cybercommunity. Additionally, the use of many native languages has only confused the African and has further distanced him from members of other speech groups. However, Negritude scholar Leopold Senghor said that the languages of Negroes (sub-Saharan region) have made their civilizations different from those in Northern Africa, where foreign influence was limited. Hence, if sub-Saharan people successfully develop a few of their non-European *lingua francae*, they might be able to operate efficiently on the information superhighway.

Cosmologies

The differing views of the cosmos make Africans important targets for an information revolution because it can help in the promotion of inquiry between northern Africans and sub-Saharan Africans and among internet users. The Arabo Barbers have used Islam to construct their cosmology and to help themselves understand the world. This belief system brings order and holism to communication among Arabic users. But since Arabic traditions differ fundamentally from Abrahamic (biblical) religions in terms of spirituality, Arabs would respond differently to foreign composites for change like technology and information systems while Sub-Saharan Africans would be more responsive to new technology than northern Africans because they have had a larger exposure to European values.

Ecological systems

These have a considerable impact on culture. Life in Northern Africa is dominated by agropastoralism and rurality. Farming and nomadic activity underpin the values; nomads move to southern regions seeking pastures for their cattle. During this process, they learn new diction, forms of expression, and idiosyncrasy from the people they encounter. Many inhabitants of the sub-Saharan region are also semi-pastoralists. They are mainly Fulani, Hausa, Tikari, Masai, and pygmies living in the

Sahel Savannah and forests, and they rear livestock and do agricultural work.

Another group are semi-urban and urban dwellers whose chief ecologico-social activities include trading, tailoring, and craft design. Urban dwellers obviously have Western-oriented professions that include medicine, law, teaching, and public administration. It is important to add that both groups--semi urban and urban-- interact together.

Colonialism

Theories of colonialism usually describe the term along dependency lines. Colonialism refers to the exercise of complete authority over one system by another. In Third World studies, the term is frequently used to delineate exploitation of a country by a foreign, usually European, country. Exploitation involves the extraction of mental and physical resources, both of which were used to improve industries in European industrial areas. The exploitation of mental resources is arguably more intensive because it involves brainwashing—making a person see life entirely through the dictates of the exploiter. If people are not brainwashed, they can preserve their physical resources and can eventually use them to improve their minds and their environment. Guinea-Bissau's foremost revolutionary leader, Amilcar Cabral described colonialism on two levels: "lies and violence. The former was a cause-effect device of unhonored treaties and unfilled promises, and the latter a power structure designed to impart fear and terror in the minds of Africans. Cabral believed that colonialism only became effective when colonizers used superior technology. Technology transformed means of production, upgraded work ethic and raised the standard of living among many Africans. Nyang (1995, p. 19), has said such a revolution created a world market which linked the human components of technological activities in the industrial regions of the Northern Hemisphere and the agricultural colonies from where raw materials were acquired and transported to the colonizers' industrial centers. That dependency-creating practice of using agricultural centers to extract raw materials may be raised to a new level with the coming of an information highway in Africa. This time, the purpose will not be to extract raw materials, but to invade the psyche of African IT users. But Cabral depicted colonialism in a negatively positive way. While colonialism led to the extraction and exportation of Africa's resources, its dialectical development provoked a new mode of production which could help African-based business compete for better markets globally. The introduction of new techniques of production and technology like coffee-extraction machines, railway lines that connected cities and made travel easier, and telephone lines, though intended to facilitate life for the foreigners in Africa, created a new incentive—a higher rationale for competition among the natives. Although there was competition in precolonial societies, it was limited. Since African societies were largely patriarchal, possession of wealth was mainly based on lineal inheritance with power transferred from father to son. Women did not acquire wealth, except in matrilineal societies in Ghana, Senegal, and Mauritania. Unless a father or uncle gave a man a piece of land to build a homestead or do agropastoral work, the man had no land. Unlike land tenure procedure which was determined by inheritance and by philanthropy, wealth was measured by the number of women and livestock a man possessed. No man could have more than one wife or goats without being given enough capital by his father. However, through colonialism and religious movements, most of those practices changed. Police officers selected from among villagers and trained by colonial administrators intimidated the latter with guns and even used brutal force on everyone, including traditional leaders whom the people revered. Through this practice, colonial administrators were able to destabilize indigenous administrative and economic structures. Another by-product of this destabilization was that everyone was given a chance to seek new ways of improving individual living standards. Parents were able to allow their children to acquire a European education in school, which eventually helped them to get better paying jobs. This new class of people was able to purchase land, live separately from the poor, and provide employment to their own kinsfolk. Additionally, people were free to choose a religion. Many converted from heathenism to Christianity and were, thus, able to receive additional education, appear more enlightened or superior to their kinsfolk, and further raise their standard of living.

Other religions can be credited for laying the groundwork for the establishment of a free market enterprise, by introducing an alternate political system. Islam, which spread mostly across central and northern Africa, changed the course of authority by influencing and altering property rights; wealth acquisition and transfer of inheritance. People loyal to women with the guile and softness often attributed to the latter, became subservient to men's authority—frequently identified with valor and force.

It is not only the change of power that accounts for the proliferation of new ideas, but the process of transferring power. Power keepers and those subservient to it are reluctant to change because they fear the unknown. A system of behavior cannot change immediately, because it involves mental activity and a certain degree of control. The mind responds to a new system through stimulus and then places it in its subconscious box until a similar system reoccurs before the mind can prompt a reaction. Like colonialism, it has taken decades to change the modern values of Africans. Basically, values changed at the levels of aesthetics. Perceptions of beauty and the liking for Western music, clothes, and language spread from the townships where natives had more contact with Europeans, to the villages. Although natives were slower than township residents in inculcating colonial ideas, including Christianity and Islam because of their deep involvement in ancestral ways,their philosophical abilities were eventually suppressed by the newness and attractiveness of foreign systems of thought. Those systems—Christianity, colonialism, and Islam—set the pace for the enactment of the industrial and information revolutions in Africa, by giving Africans a new way of seeing life. But Africa has not fully experienced an industrial revolution because of its "inferior" products, imperialism, an unstable political climate, a stronger foreign influence on indigenous political activities and other foibles. The willingness to elevate individual living standards and enhance economic skills did not end with the departure of the colonizers and religious imperialists. Such conditions prevail in the semi-urban, urban and rural communities and telecommunication providers should become aware of them. Certainly, there can be no sustainable degree of awareness without an understanding of how most Africans think.

The African Mind

Because of a neopatrimonial culture, where power sharing is personalized economic logic has become hostage to political logic. The professional capabilities of the public sector and the legitimacy of private and public organizations have been weakened by bad government policies, so has government's crack down on competing political forces caused a fluctuating economy. New civil societies have emerged as a result of migration, but they have not been able to play any significant role in national economic development or governance. Additionally, legal and regulatory bodies created without a democratic process have not promoted private sector confidence in the government. This distrust-oriented state of the African mind has caused severe constraints in strengthening human capacity building efforts, at least, as far as investing in human capital, institutions, and programs.

Communicating at the interpersonal level is universal, although African tribes and kingdoms have always placed special premium on interpersonal communication. Dialogue, an integral part of interpersonal discourse, constructs wholesomeness; it makes Africans to better understand of their environment and to relate to it. Dialogue also controls Africa's socioeconomic paradigms. Ideas and activities are negotiated primarily through dialogue, despite the presence of several languages in any given village community. Children are limited in what they tell adults and what they need to know about adults. Information regarding personal medical records, sex, and bank accounts may be shared between parents and children in Western societies; however, African parents do not normally disclose information on such matters for fear of breaking down the barriers of allegiance their children owe them. For parents, such information is strictly privileged. Classified or privileged information make communication in traditional sectors an elitist treasure. The hopes and fears residents adopt and identify with, how they think, what they think about, their instincts, whatever stimulates them to act or react, their reverential treatment of the cosmos, fauna and flora, and factors that motivate them to change, or adjust their thoughts and actions to suit their needs, are all holistic constructs. Members of secret societies like the FBI, CIA, Scotland Yard and the Kremlin keep classified information among their own ranks, which can be used for or against members of the public. This communal *esprit* constitutes heavy communication activity in agrarian regions.

At the administrative level communications is an incidental concern. I Ngwainmbi (1991, 1995a, 1995b) detected very limited interaction between government workers, who make up 85 percent of wage earners in Africa, and civilians, who make up 80 percent of the entire population. Government personnel in such professions as: education, health, information, and territorial planning have failed to effectively communicate administrative and public interest information. Additionally, most Africans earn terminal degrees in foreign languages, hence their reasoning or psychic stimuli favor European culture, not that of Africa. This pathological zenofelea is partially responsible for the negative events in Africa, whose policy makers use mainly foreign values in designing and implementing public policy. An African Ph.D. degree holder is more likely to speak a western language among his kinsfolk than the language his people speak, in order to appear distant, strange, different—superior. Because most Africans do not question authority— a tradition extending from ancient Kemetic monarchies—foreign-trained government workers consider communication with their rural kinsfolk unnecessary. Hence, the reason most government-sponsored projects are not sustained.

The absence of effective interaction among government officials, and between them and civilians, accounts for non-accountability in the use of public funds. Ancient structures which involved interaction between the governors and the governed (Ngwainmbi, 1995a, Chapters 2 and 3), but which were destroyed during colonial administration, remain dysfunctional in postcolonial Africa, partially because native Africans with European education control the economy and have not been able to reconnect with their people. This elite-mass gap has reinforced the psychological damage initiated by colonial administration. Without effective interaction between the elite--educated ones--and the people, there can be no cohesive action and no comprehensive context for gathering, analyzing, and implementing progressive ideas. People seek political office in order to satisfy certain needs which may include generating respect from, or instilling fear in, others. They use such power to embezzle public funds. Incidentally, no post-colonial African leader was known to be rich, prior to taking office. The struggle for political leadership in Africa, then, has been motivated by the need for an improvement in individual economic and social status. As a result of a change in personal need, the desire for self improvement has become a greedy one, but its fulfillment is limited in time, space, and quantity. That need must be shared for self fulfillment to be maximized.

Political activities are generated through the drive for self-assertiveness, self-satisfaction, self-identification and self-expression. A bad economy may accelerate hostile political activity between the haves and the have-nots, with the latter seeking to become the former. The have-nots strive to satisfy their needs by criticizing the haves. However, there is no guarantee that such critics would correct the evils for which they criticized the power-keepers when they gained power. It is important, then, for critics to contribute ideas in order to improve local economy. In so doing, they would be reducing greed, which has always produced bad leadership.

A functional political process is supplemented by a stable economy. Both factors can only be sustained through free and frequent interaction between members of all social classes. Hence, the future of any human community lies not in its political, military or technological strengths, but in the ethics that produce incentives for identifying and pursuing them. What a community adopts in time and space, its hopes and fears, its worldview, how it considers its failures and successes, what it thinks about, how it relates to other persons and things, what causes it to act or react, how it perceives the cosmos, fauna and flora, how it uses its resources and changes its thoughts and actions to meet its needs, constitute its morals—its values.

Cognizant of those factors, it is appropriate to stipulate that Africa is not prepared to influence the course of the information revolution because it has no significant role in the fast-changing world where technology is making information available to everyone with access to a telecommunication product. Would participating in this supermarket mean the end of privacy or confidentiality which Africans revere? What role would African governments play? Who would direct the superhighway? How would national and individual security be protected when cyberterrorists invade cyberspace? Would newspapers, television station networks and postal agencies which governments control become obsolete? If so, how would African ideologues and ideocrats influence other peoples' minds? How would they use cybertechnology to consolidate power? Or would the influx and availability of cybermediated information mark the end of dictatorship, making people cease to be dependent on government-sponsored media for public interest messages? How many Africans would participate in this new world order when 85 percent of its population live below poverty level (according to UN standards), and when less than one percent have personal computers? In order to answer these questions satisfactorily, providers and exporters should understand that the African is consumer-oriented rather than product-oriented. Through their holistic culture, an integral part of which is sharing, they have not been able to develop effective economic skills. Workers use office time to host friends and family members, as being officious during such visits might be interpreted as disrespect for family values. In semi-urban and rural areas, it is normal for family members, friends, and acquaintances to "borrow" goods from a store keeper who is their kinsfolk—goods they never pay for. The store owner does not refuse such favors for it would be regarded as greed. Like everyone, he is expected to commune—share his wealth with his kinsfolk. Most store keepers eventually lose their businesses by lending or giving away their goods to relatives. Hence the potential for maintaining a steady market for foreign goods is limited.

Strategies for Improving Services

In order to reduce the level of dependency on local business managers, providers of the information technology should impose foreign capitalist values on local sales agents—no payment, no service. Local business partners should often bring foreign sales managers into the semi-urban and rural business areas because they only understand business language. The international telecommunication provider (a native or foreigner) should establish trust with local consumer markets by being consistent. In other words, prices of information products should not fluctuate frequently, store hours should be kept, and products should be effective. Information on labels should be legible and understandable by the consumer. Some foreign brand products should be labeled using local language. Retailers should also understand that locals are often skeptical of international business transactions for the following reasons:

1. They cannot determine the validity of business;
2. They are reluctant to spend on foreign products;
3. They are accustomed to a corrupt internal system—full of dishonored promises, fake products, and unfaithful sales people. Some salespeople collect money but fail to deliver the product;
4. Longer delivery times.

The African Investor

International investors should also know that locals would be reluctant to invest in, or obtain international business for reasons 1 through 5, and like the retailer, he should therefore take the necessary steps to establish long-term relationships with local investors. On-site showcasing is also necessary. Since seeing and/or experiencing phenomena enhance belief, foreign providers and investors would increase the percentage of African investors by showcasing their products in African communities. It is advisable for a foreign provider to invite qualified prospects and potential investors in Africa to a showcase, even if the cost of setting up the showcase is higher in the target country. Generally, market testing done in foreign, especially European countries have not produced positive results for the African consumer because samples used were unrelated to realities of the African environment. For example, a telephone market in Geneva will be different from that in Senegal because prospects in Geneva have been exposed to telephone sets and telephone services for a much longer period than Senegalese. Moreover, African-based qualifiers would be pleased to know that a foreign company respects their purchasing potential well enough to showcase products and services among them. For African investors and consumers to make informed decisions regarding the purchase and reception of telecommunication products, the products should be marketed to them, with their unique needs and applications in mind. In addition, prospects should be exposed to the products.

Information Technology

The operational meaning of information technology, (IT) provokes a critical question: How should countries with limited infrastructure understand and master an information economy? Understood as the science of collecting, processing, storing and transmitting information through telephone lines, teletext, satellite, cable, and facsimile, IT performs major functions, especially in developed societies, including the promotion of social and economic development. It is used as part of the infrastructure for strategic competition and management activities and to deliver goods. In most developed countries, television sets, cars, telephones, refrigerators, calculators, photocopying equipment, cash registers, optical fiber transmissions and automatic teller and cash machines are operated by computers. Money is sent millions of miles through computers. Computer companies are now IT companies. There is no doubt that computers are largely responsible for the new IT revolution. However, computers do not have the capacity to learn, translate, articulate, or use the five senses that make up the human intellect. Therefore, mankind must not rely solely on them in order to function. Although documentation is delivered with each computer system, Africans to whom this phenomenon is a novelty, will nevertheless encounter serious problems. Africans think holistically, not in the metrical pattern which typifies computer operation.

Cost is yet another reason why despite its enormous potential to create, keep, and spread information, the computer will remain out of reach for most Africans . An average computer costs about $2,000 which is equivalent to millions in most African currencies. Only upper-class workers and business executives can afford its cost. Thus, the primary beneficiaries of this technology will be the rich. How then should the less fortunate tap into this information supermarket? Davies (1985) has the answer:

1. An analysis of the actual information needs of the country should be taken.
2. A realistic appraisal of the existing information infrastructure is needed.
3. Projects should be started at a local level and (providers should) concentrate on the capture of local information resources.
4. The experience and control developed in these small scale projects can then be integrated into national and regional information systems.
5. Software and hardware which can be used for direct access to relevant, available and appropriate information must be developed, rather than merely providing rapid access to citation.
6. Information networks which can rely on extensive use of semi-automated and independent information processing as well as fully automated, linked computer systems must be developed.
7. The employment of information workers such as librarians and other information professionals to act as intermediaries between advanced information technology and information users will aid in the delivery of information in a form that is appropriate to the user (p. 254).

Potential Benefits of IT to the African Economy

Although these options were identified over a decade ago, they are still extremely relevant. They remain critical to information utilization in developing countries and could be used to the advantage of the countries. Because public records are kept manually, civil servants manipulate information, destroy documents and even negotiate bribes with members of the public. The micro-processor could be useful in curbing such practices, by requiring standard procedures for logging in basic data like names, birthdates, and money, and amounts of local taxes. A computerized information storing system would make public records easier to be monitored, stored, and used upon demand. However, the absence or inadequacy of computers in government offices makes the reliability of documents highly suspicious. People can obtain more than one birth certificate, identification card and qualifications are routinely faked. This amount of fraud has worsened Africa's economy causing continent to depend heavily dependent on foreign assistance.

Desptie that condition, IT is a development agent in some African countries, though on a small scale. Some countries are using databases and computerized information systems for traffic safety activities, road networks, and billing and accounting systems for transport investment, planning and management of railway systems. For example, the Ministry of Finance in Morocco has a computerized auditing, control and tax administration system. More computer use could curb both the massive rate of fraud in manually processed auditing, and the 60 percent poverty rate in Africa that has been caused by public funds embezzlement.

CD-ROM

CR-ROMS are yet to become available for the general public in African countries. Used elsewhere to retrieve entertainment messages, CD-ROMS are needed by researchers in Africa. They were only introduced in African in early 1990s. ExtraMED, a full-text database of biomedical journals and illustrations available on CD-ROM was started in 1994 to store new information on tropical medicine and tropical diseases like malaria, cholera and other water-borne ailments. Yet, according to *CD-*

ROM for Development, 1994, a newsletter produced by the American Association for the Advancement of Sciences (AAAS), the information available to Third World journals through ExtraMed is not supplied by MEDline—the world's largest medical database. While some may be flattered by the existence of an African-based electronic medium that provides health information worldwide, the source of this information is unclear. No records have been available to indicate the exact number of local subscribers. As for MEDline, the recognized, well-respected database used by developed nations, an annual subscription fee has been running about $3,000—equivalent to four month's salary for a medical doctor in Cameroon. Only medical professionals in industrialized countries can sustain the operation of this information source. The elitist nature of MEDline and other computer-mediated communication make the toll fare for the information superhighway even much higher for poorer countries. This further alienates them from exploiting the advantages the technology offers.

The Information Superhighway

Since the production of the first mass communication agent—the bible—the human being has continually sought ways to relay messages to mass audiences. Man is constantly in search of more effective ways of staying in touch with fellow man twenty-four hours per day. Telephones, beepers, computers, fax machines, and other systems have made life more bearable. A cursory examination of human behavior reports its insatiable quest for information. Human beings have a history of gathering, processing and disseminating information for profit, knowledge acquisition, or entertainment. In closed societies, information is not only used for entertainment and knowledge-gaining purposes. Tribal leaders in ancient times, gathered information to foster community spirit and to consolidate their authority, dynasty and territory. Information has also been passed from one generation to another, to maintain power and tradition.

In modern society, information is created and used to measure the performance of an economy, to persuade, misinform, maneuver, improve standards, and promote education. The channels used in the management of information continue to change along with the increasing need for information. Information is either classified (privatized) or unclassified (publicized). Information designed for public consumption is channeled through long-distance means: television, radio, drum, bell, flute, gun-shot, siren, and satellite. Classified information is channeled through private dialogue, telephone, fax, beeper, caller ID, sign language and telephone. Although third parties can interfere with their use information is valuable and sensitive.

An information highway has the romantic connotation of an open road stretching beyond a horizon flanked by electric and telephone poles. Patton (1986) found the information highway to be:

> the most expensive and elaborate public works of all time offer(ing) a vision of social and economic engineering. It was planned to be at once a keynesian economic driver and a geographic equalizer, an instrument for present prosperity and the armature of a vision of the future (p. 17).

Not exactly. The brainchild of technologically advanced societies and larger economies, the information superhighway (IS) is a combination of Internet and metanet in computer networks linked to powerful military bases, research centers, universities, and individual private homes that provide a new economic and social direction for the human society. Promoted by American political leaders, the information superhighway (IS) creates a new economic and political order. Some critics are concerned that it is a ploy by the U.S. and other industrialized societies to consolidate their authority over the rest of the world. This political dimension of the term implies that the IS is a program which when fully implemented, would solidify America's economic and political influence over world regions.

Another perspective on the program is that it promises a renewed sense of community and in many instances, new types and formations of community.Jones (1995) thinks computer-mediated communication would be achieved through electronic pathways what could not be achieved by other forms of interaction—it would connect people, rather than atomize them. But the industrial revolution separated people from one another. Machines' replacement of human labor, the quest for individual material wealth, and the presence of tarmac roads, cars, airplanes, telephones, and trains have psychologically detached members of the human community from each other. This way of life has reduced human beings to robots and has made life more complicated, more absurd, and less exciting. People are not as friendly as they used to be before the revolution. There is intense competition for everything but spiritual satisfaction and mental sanity, the real attributes of happiness. But an IS that promises a new community spirit through electronic communication only satisfies those wealthy enough to afford its cost literally and metaphorically. Reingold (1993) has seen it as "a place where individuals shape their own community by choosing other communities to belong to" (pp.57-58).This condition only enhances the very capitalist practices that separate the rich from the poor. Policymakers and Internet subscribers are trying to change this. Phrased differently, any computer-mediated communication network member will best interact only with those linked to his/her computer. He or she will generate, send, receive, and share information primarily with those connected to the same network, thereby creating a separate elite community from the mainstream community. Also, the idea of choosing from a menu what message to retrieve or send makes the functioning of this computer-mediated community a paradox. Community members live together in the psychophysical world, but when membership becomes a matter of subscribing to an electronic information cluster, the interaction level is bound to be different This neo-Marxist culture redefines class where the superstructure—the rich ones—have privileged access to information, while those outside that community only rely on cheap and often polarized channels: television, telephone, word-of-mouth, and public gatherings, for important information. If electronic communication enhances socioeconomic coercion, why should members of that community choose who to communicate with? Why should their skills be limited to an available menu? For how long will members remain loyal to this new community?

The quest for electronic information is no longer a dream; more business and research institutions around the world are subscribing to the Internet. However, the poor masses may never see this highway, even if information about it is passed on to them. Citizens in developing countries may benefit from the technology only when:

1. Its primary users, civil servants, health and business managers, interpret and share the information received from respective experts connected to the Internet;
2. Those entities collect development information from underprivileged ones in an organized fashion, and circulate it to other members, in order to supply information for the group;

3. Policy makers in poor nations can use the Internet to create contacts with major corporations and strengthen diplomatic links with major world leaders connected to it, not only for entertainment;
4. The Internet can be used to help curtail the costs of traveling abroad for political and economic conferences, although false computer messages can also thwart the success of this practice.

National Dilemmas

The transmission of economic data through new technologies will enrich multinational corporations at the expense of developing countries. These corporations have sophisticated infosystems and trained personnel (economists, computer experts, and communication consultants) whose services they can use at their convenience, to exploit information and material resources from poor countries. Schiller (1970) predicted that:

> Poor nations and voiceless sub-groups within countries, developed and non-developed, are and will probably be shut out from these powerful new capabilities of administration and governance. Unless there is social mobilization and awareness not now apparent, further domination and dependency will be the likely accompaniments of the extension of the new information technology (p. 30).

Those predictions can be supported by the following facts:

1. Developing countries have limited technology to harvest electronic information, hence, they cannot influence world politics and world cultures;
2. These countries have no political power to influence powerful nations;
3. Much communication activity is obtained mainly through hearsay with limited effort from authorities to create fast communication channels. This lack of zeal has been enhanced by dictatorial techniques;
4. There is less emphasis on exporting commercial resources,

hence a lesser incentive for inventing and utilizing telecommunication products;

5. The importation of such technology may result in importing experts to train indigenes. This personnel usually bring taboo behaviors from their culture that infiltrate the elitist lifestyle of the indigenes.
6. Imported technology reduces local incentive and eventually replaces indigenous technology;
7. The greatest concern for patriots is, they would lose their native consciousness when the information superhighway passes through Africa. The polarization of indigenous culture appears to be no one's concern, certainly not that of business executives in industrialized nations who only think profit.

For a society that practices secrecy, sacredness, and communalism, any new system of behavior would:

1. Threaten its ancestral roots;
2. Lower its unique qualities by spreading a pluralistic way of living;
3. Create an identity crisis between generations by introducing a new way of seeing and reacting;
4. Eventually diminish the importance of African languages which carry afrocentric wisdom.

Most Africans will lose their modus operandi and inculcate a computer language that has been designed basically for the European psyche—for those who think primarily in metrical terms. Nevertheless, steps can be taken to avoid those problems. They have been provided in the latter section of the chapter.

Indigenous Habits and Foreign Corporations

Telecommunication corporations, pharmaceutical companies, and other economic forces have been aggressively seeking and establishing new markets in Africa. At various international business meetings, economists see Africa as the next economic superpower, which hitherto has been ignored globally because of its dubious governments. This unprecedented interest in Africa leaves some second guessing, especially those who believe that nature, from its origin, had condemned Africans to kill and perish from famine, like the Biblical Sodom and Gomorra. An examination of human history shows a trend in race progress and race retrogression. Africa has undergone massive torture from its own and from others—the progenitors of something called "modem technology". Africa has witnessed centuries of exploitation, from the hands of missionaries, colonial administrators, and the UN Security Council, which negotiates control of economically dependent world regions through industrialized member countries.

Is it right to declare the 21st century the era of techno-political imperialism for countries in the Southern Hemisphere, when countries from the Northern Hemisphere with advanced communication systems offer fiber-optic equipment and manpower in order to facilitate access to global information and to make a profit? Offering "gifts" to African tribal chiefs in change for friendship, cheap labor, land, and local slaves centuries ago is no different from cajoling African policymakers, through diplomatic and business meetings outside Africa, into implementing foreign operational standards for local telecommunication operations. Africanists have suspected the latest plans by industrialized nations to subsidize sponsorship of African telecommunication infrastructure; however, they cannot halt the projects that these entities have been implementing for decades, nor do they have the power to pacify the ideology that bringing modern technology to Africa is most appropriate for Africa's economic growth.

Conversely, the search for the end of dependence by developing countries may soon be over with the importation of modern telecommunication products if the latter are used correctly, Africans would be able to understand and relate to people from other cultures in an enlightened way. Moreover, the users of over 800 indigenous languages in Africa have not been able to improve understanding among themselves; their languages have only helped to confuse the people. That is why any opportunity for conflict between tribes or nations leads to war. African Romanticists who promote patriotic values do not have media technology to resolve this problem and promote an African-centered agenda that can help them resist the abundant influx of European ideas and international technology. This inability leaves feeble African leaders no other choices than accept offers from international institutions.

But why should they settle for a fair share? How do those kinds of negotiations help the African people? People immune to dictatorship in Africa would expect dictatorial regimes to pay for cable installation, and control cable operation. As stated in Chapter 6, African telecommunication officials would not implement policies that promote equal access to data regarding government jobs, education, health services, law, or any topic dealing with the privileges of major government officials, since the immediate availability of such information may expose their corruption. Because African leaders own and control the mass media, they are likely to closely monitor any cybermediated information flow, and/or importation of mass media products.

Since democracy is a new phenomenon for these nations, evolving electronic devices and services should be deployed to meet the public's needs, desires and interests. Moreover, in a democratic system, every citizen has the right to know, and deserves to have access to a wide range of information services. Fiddler (1995), and Dordick (1995) also assert that a national information service is necessary to meet the information needs of all people.

How can a functional information highway affect local communities, governments, business organizations, other units of society, and democracy, if all citizens have equal access to a wide range of information? Africa's major policymakers cannot accept both democracy and new communication systems wholeheartedly, since their citizens lack confidence in them and in the military. In addition, an active information market could be dangerous because too much information destroys, causes confusion, and affects decision-making in a negative way. Already, some Internet users in America have been victimized by the product. The Internet has given the user direct access to information and has created serious moral, ethical and policy problems: children have direct access to cybersex, which contains profane sex language and sexually explicit activity. Anyone having access to the internet can see cybersex for monthly fee of $30 or less. CNN, an international TV news network, otfen reports that people disappear into cyberspace after meeting strangers on the Internet.

Although the use of that service is economically beneficial to the computer company, it cannot control the message content or implement successfully any policies that negate the free use of the internet, because that would mean violating the First Amendment—a law which gives American residents the right to free speech. This problem would be more prevalent among African Internet users whose leaders neither have the sophisticated investigation capabilities of the FBI to track down libelous internet users, nor the discipline to consider any computer-mediated messages hazardous to their citizen's well-being. Although African governors successfully prevent the use and circulation of most foreign media containing taboo information, like *Playboy*, pornographic video cassettes, and newspapers with radical ideas, they do not have the technical capabilities to control the content and circulation of Internet messages transmitted interpersonally from one residential computer to another.

Despite those cons, policymakers can use the new information highway to foster unity and progress. They must consider policies that help close the gap between them and the public, by creating a database that provides citizens with free access to information on:

1. Healthcare facilities and disease prevention measures, local addresses, phone numbers and locations of hospitals, clinics, doctors, nurses and pharmacies. The same information on traditional healers and herbalists should be included in that database;
2. Databases should contain information on schools, tuition, degree programs, local administrative offices, their operation hours, and services they render. Information on the voting process and political candidates, basic legal issues and individual voting rights and privileges should be in the computer.
3. Information on commercial activities and business contacts, e.g., access to local markets.
4. Information specialists should create and broadcast distance learning programs and conferences to international audiences through satellite.
5. Media practitioners should be hired to operate these systems in different regions of each country.

With the technology, the government of each country should also be able to:

1. Facilitate communication among citizens and bring government

closer to people, and between civil servants, and foreign business, social and political institutions;

2. Expedite the downsizing of government and reduce its high spending rate. More people should have access to a wide range of information sources. They will be able to cultivate creative and self-reliant skills and seek employment in sectors other than government. This will:
   a. increase per capita income by enabling civil servants, who make up approximately 80 percent of each country's workforce, to seek jobs in the private sector;
   b. reduce the unprecedented heavy dependence on government. People would then have more control over their destiny,
   c. increase the potential for middle-income jobs and a free-market economy;
   d. Encourage free individual enterprise by making people to think for themselves and find creative ways to earn livings

Technology versus Performance

Ignorance in using technological products automatically induces an inferior feeling, indeed, a sense of loss. The rapid development of technology reflects mankind's desire to overcome ignorance and improve living standards. The developers of technology have been empowered to control other human groups in civil society. Telecommunication institutions control other institutions at the economic level. Mowlana (1994) has succinctly said:

> The magnitude of the economic importance of the telecommunication sectors of the industrialized countries lies not merely in quantitative overall economic output, but in the technical properties of telecommunications, in its rapid development and proliferation, and in its significant impact on other economic sectors such as banking, finance, retailing, and transportation (p. 163).

Mowlana also mentions strategic security and military aspects of telecommunication as keepers of information hegemony. The convergence of telephone and computer communications has created new opportunities for national administrations and globalization of the facilities and services by a few telecommunication companies. The packaging and transportation of information through electronic channels presupposes its economic significance to both technologically advanced and unadvanced societies.

Technology and information products are composites for evaluating new markets and consumer interests. Technology is a cause and effect system which leads to greater foreign investment. It also leads to the loss of control in the domestic economy and the introduction of foreign patterns of consumption (Stover, 1984, p. 74). This enclave economy creates, in Francis Steward's words, "a society in the image of advanced countries, requiring further imports of technology to survive and grow" (1977, p. 138). There is no doubt that African economies have struggled since African countries first engaged in trade relations with their former colonizers. Senegal, Cameroon, Cote D'Ivoire, and other Francophone nations have since independence operated solely through French diplomacy. Not only has information on world trade and other global issues been disseminated primarily through the French press agency—Agence France Presse Intemationale, with modifications to suit French taste, France benefits from telecommunication transactions between foreign corporations and Francophone consumers. A French company would sooner have a monopoly over the cellular phone market in Francophone Africa than allow access to any other private corporation, even when the latter can provide cheaper rates and a higher quality of service. Although the French are not known globally for promoting a free-market economy, they have a reputation of doing business mainly through governments. The French telephone exchange service has been earning billions of dollars annually for connecting calls from other regions to Africa, also between French-speaking African countries. However, the telephone exchange offices in Africa only receive a small share of the profit. Conversely, since African-based telephone systems are not as sophisticated as French-based systems, the African market depends on the latter for quality service. Any foreign corporation seeking telecommunication business with a Francophone country would likely negotiate for profit from French telecommunication companies.

One cannot deny the positive effects of this technology in the enhancement of living conditions. Popular agents like television and radio have mobilized forces in communities faced with epidemics, poverty and other social problems. The makers of technology enjoy conveniences provided by fax machines, high definition television, cellular phones, and multi media and the ability to receive a message anywhere by touching a dial has made communication easier and more accessible. These advances have become a luxury for millions of people in industrialized nations, who have in turn developed a desire to help foreign consumers to explore their objectives. The technology has also proved intrusive to some consumers. This group believes the preservation of their values and economy are at risk with the importation of communication technology. Not so with the North American Free Trade Agreement, NAFTA. Supporters of NAFTA, which includes the US, Canada, and Mexico, as well as the General Agreement on Tariffs and Trade, maintain that expanded trade will bring growth for skilled workers in export countries. While new export platforms in poorer nations may pose a short-term threat to workers in industrialized nations, growth led by exports would eventually turn poorer nations into new markets. However, as mentioned, importing nations may become subject to the mercy of the exporter in matters of supply, pricing, etc. When a company has a monopoly over certain markets, it can place higher prices for its products in the importing economy, while its importers suffer inflation. Such is the case with many African countries that suffer trade deficits and remain forever indebted to their donor—IMF, World Bank, CIDA, and the money-lending countries. These countries find it difficult to withdraw from their partners; richer donor countries tend to exert undue control over the diplomatic activities of their debtors. It is not surprising that the U.S., for instance, would withdraw aid to Libya, Cuba and any regime, whose leadership the U.S. does not like. The U.S. has tried repeatedly to influence China's policies over the years. The only reason it has not is, it cannot afford to forfeit access to the huge China market. Moreover, the US-China Business Council estimated that the telecommunication equipment market in China would be worth $30 billion in the next five years.

Given that trend, how can terrorist governments benefit from trade activities with non-terrorist regimes? A consistent wave of political upheavals; the killing of civil rights leaders, the denying of jobs to opposition party members, witchcraft, tribal conflicts, sycophancy, nepotism, tribalism, favoritism, sexism, and other ethical problems in developing countries indicate a bleak future for free-market economies and the slow development of a long-term information supermarket. No sooner will a state-of-the-art telecommunication system be installed when corrupt officials will attempt to restrict access to privileged ones—friends, family members, bribe-givers, and the wealthy—with limited experience. Hence, an education on the use of telecommunications for national development is necessary.

Education as a Communication Paradigm

Education produces knowledge and knowledge influences action. While traditional education requires students to be in class, listen to instructors, and take notes, students can become *tele*-learners through advancements in technology. Students can learn wherever there are wireless, micro-supercomputers accessing information worldwide. With *tele*-learning, classroom teaching will also increase dramatically. However, the need for personal contact may diminish and the university may become obsolete. And anyone will be able to tap into any classroom lecture or educational course in the world, from their living room using the teleputer (DiSalvo, 1995, pp. 18-19). This means that people with access to a teleputer, irrespective of their enrollment status, or their cultural and educational background, sex, creed, religion, or age can, in addition to listening, participate in the lecture. If that individual is not familiar with the content or level of the lesson, or if he/she does not support the subject matter being taught because it conflicts with his/her values, the individual might disrupt the session. The capital argument here is, the tapping of any classroom lecture is piracy. Having direct access to a classroom session being taught in Africa is likened to secretly watching a sexual activity. Although it may help the self-proclaimed monarchs of knowledge (African professors) to better prepare their lessons and augment their teaching standards, tele-teaching may violate the privacy and concentration that learning requires. It will also violate the African concept, that conventional wisdom is a sacred entity that is passed only to people with special qualities; benevolence and respect for elders. Because traditional Africans consider the sharing and spreading of public-interest information the sole privilege and responsibility of village leaders, title holders, messengers and other authorities, the proposed practice of tapping into computerized sessions may distort the organized pattern of packaging and distributing public-interest information. It means that anyone using a teleputer is capable of participating in the generation and processing of classroom information. That individual has the same accessibility to watching, listening, and speaking back during the session. This may startle many African classroom instructors educated under the "school of hard knocks"—a utilitarian philosophy, reminiscent of the British educational system reflected in Charles Dickens *Hard Times,* where learning was absolutely useful, not entertaining. For instructors in this school, teaching is an elitist profession, not just another job.

The 21st century may be the age of enlightenment around the world, as more people having direct access to electronic information systems would know more about other cultures. Certainly, international and interpersonal relations will improve drastically, as more people would have equal access to different perspectives and different ways of life. With improved international relations, the potential for non-implicate diplomacy, better business transactions and economic redevelopment in underdeveloped nations will be high. The power of dictators will be neutralized as classified information will become available. Dictators will no longer be able to plan *coup d'etats* in secret or keep such information under tight control. The information superhighway will give rise to democracy. There will be more free speech. By interacting on the Internet, people will be able to exercise their rights to know, judge, and to act on a common agenda.

Communication technology will heavily change the way people learn and behave, as future generations will have more access to worldwide learning resources. After all, more knowledge brings more wisdom which helps in the making of better decisions. Better or informed decisions bring satisfaction—happiness. Despite its advantages to humans, the information technology will be more useful in European societies where privacy is not so much an issue as in Africa. In European societies, people are more open to change, more curious, and more receptive to newness than in African societies where spirituality controls life. Africans are sometimes less receptive to change because they believe a Supreme Force is in charge of their existence. Therefore, they wait for signs before they act. This holistic approach to learning is responsible for the lack of an incentive for an industrial or technological revolution. Hence, the levels of productivity in, and responsiveness to, new technology cannot be high in Africa. If Africans are to use the information supermarket for economic advancement and political stability, they must reevaluate their customs and abandon practices like witchcraft, nepotism, tribalism, and laxity in the workplace, which have only deterred progressive change.

Framework for Understanding Technology's Financial Import

When a company finds a new foreign market for its products, obtains an export patent, and is able to transport the products, the company can also increase its revenues through additional sales. The importer can use the products wisely to increase opportunities for profit. The exportation and use of communication technology is an economic redemption for more than one million jobless university graduates with degrees mainly in academia. These people, who lack the technical skills to market themselves nationally and internationally would benefit considerably from the importation of the technology. They will be trained for technical and managerial jobs by satellite, long distance telephone, fax machine, and computers.

Networking in Academic Institutions

The concern that degrees obtained from universities in underdeveloped countries are of lesser quality than those garnered in developed countries is valid, because libraries, faculty and students in underdeveloped countries do not have enough funds or information infrastructures to conduct research in their respective fields. In Francophone institutions, students regurgitate material from professors which the latter have obtained from old and out-of-print books. These professors neither have recent editions of the class textbooks, nor seek them. Most of them merely transcribe information from old books and use them to teach the same class for many years, due to a shortage of support/instructional materials like videotapes, audio tapes, manuals, and new literature. Professors also use redundant notes, most of which are their own unresearched and untested opinions, or are material from few sources. A university with about 10,000 students may have only one copying machine. Instead of operating computers to document research findings, some science departments still use paper and ink, or test tubes which are often destroyed, lost, or contaminated. Social Science professors award degrees based on theses laden with provincial references, even when the subject requires documentation of international sources. Students then pass on to higher levels of university education with half baked ideas—scanty information— in the area of study. This lack of adequate information leads to the issuance of what the international scholarly and business communities perceive as low-quality degrees, and find them less marketable internationally. However, statistically, African graduates have outscored their Eurasian and American colleagues in class and board examinations. Therefore, one should perceive students in underdeveloped countries not as inferior degree seekers but as entities with superior intelligence being supplied with inferior opportunities.

Even the New World Information and Communication Order (NWICO), which found information equity an important asset for all regions more than fifteen years ago, seems to have abandoned that position. The prices of new books and other published learning material are still too high for academic institutions in underdeveloped countries. Students enrolled in a course in an American university can purchase a reference book that costs $30. The same product equivalent to 15,000 francs, CFA cannot be afforded by students in Francophone African universities. If they purchase the book, they might not have other funds to obtain other class materials. African institutions have not had equal access to information due to their overall low income.

Information Resources

Electronic Systems

Despite these obstacles, developing countries are gradually sailing into the information supermarket through the joint efforts of some universities and international educational research organizations like USIA, CIDA, UNESCO, and the American Association. for the Advancement of Sciences (AAAS). The latter has been holding workshops with some African academic institutions on electronic networking. In 1993, the Association of African Universities (AAU), and the AAAS held several workshops in Accra, Ghana to create opportunities for West African universities, and to participate in academic benefits afforded by electronic networking technologies. It also sought to facilitate research collaboration between African professors, students, and researchers and their American counterparts. Through this effort, information could be cross-checked and researchers could have access to PC computers and effective applications. A project to sustain academic and research networking was successful in Zambia. UNZANET, an academic and research electronic network operating out of the University of Zambia, has been serving the entire country including non-government organizations. UNZANET sent its first recorded message to an African interests network in Washington, DC in September 1993. UNZANET, whose growth was assisted by the Rhodes University, South Africa, and which received an average of 70 calls from system users daily with approximately 200 kilobytes of mail and 200 messages in November 1993 (Robinson, 1993), was linked to the world internet through regular automated computer-to-computer calls. The differences between the total traffic flows for local-to-local, local-to-international, international-to-local, and international-to-international transactions between October 1991 and December 1993 showed that UNZANET would become a potential socioeconomic indicator, especially to local subscribers. This is evidenced in the 2,400 local and 90 international calls processed by UNZANET in 1994 (see Chart 4.0). During this period, local calls to UNZANET averaged 1,000 per month. This shows its increasing importance as a viable information outlet in Zambia. The larger the volume of local and international calls, the greater the potential for business transactions. International transactions were made to UNZANET through WORKnet-an NGO network in South Africa, and SAFIRE, a drought and famine relief network funded by UNZANET, whose growth was USAID. Because of financial difficulties, UNZANET was expected to be connected only to regional neighbors like Kenya and Zimbabwe through a 9600 ban dial-up link. However, its connections now include other areas within Africa and in Europe (See Figure 4.0).

With such progress, an interregional connectivity is imminent. Moreover, it would be cheaper to route e-mail through third parties like South Africa or Britain, which have a larger clientele, than through direct mail processing. Confidentiality cannot be guaranteed in this case.

PADISnet, the Pan African Development Information System, is arguably the largest network set up in Africa to advance data communications, improve information flow for economic development, and enhance the timely utilization of existing information systems. It has been used mainly for e-mail conferences and bulletin boards—which are essential for communicating development messages. Computer experts in Africa have held consultation, conferences for network promotion, to discuss

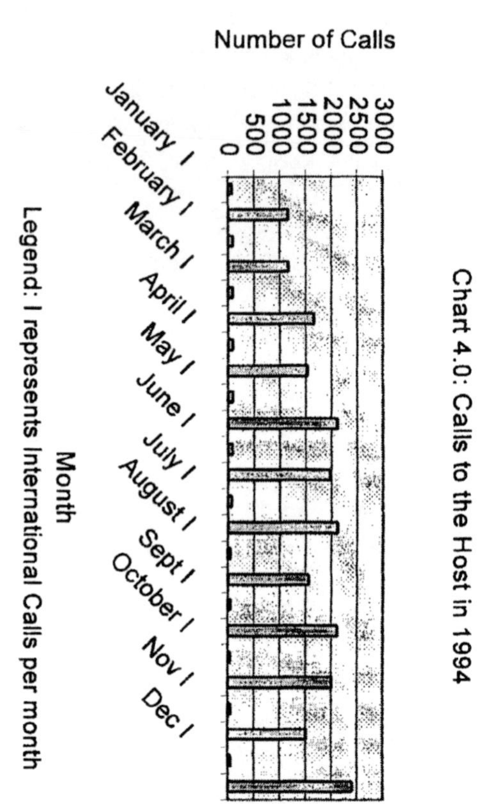

Chart 4.0: Calls to the Host in 1994

technical problems like modem configuration of software, as well as legal matters (Adams & Hafkin, 1993). However, the sustainability of such transactions is suspect due to a shortage in human resources, up-to-date information, and funding.

Figure 4.0: UNZANET's International Connections

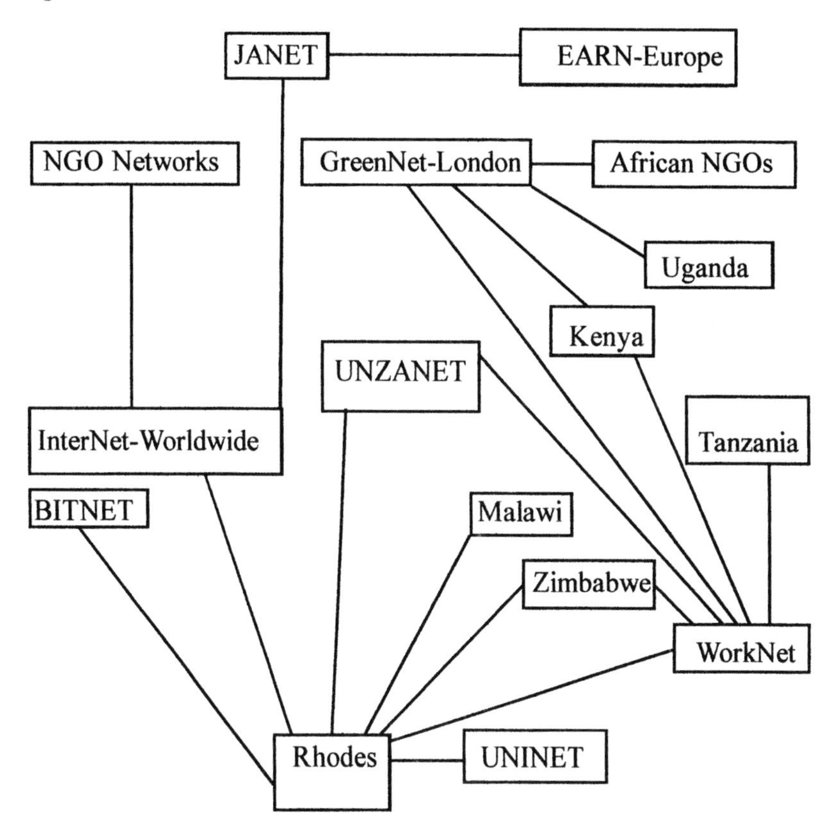

Source: *Electronic networking in Africa.* Washington, DC. AAAS.

Libraries

Most African research libraries are not well equipped; they lack computer databases. Users depend on card catalogues, some of which are torn, or displaced, in identifying books and other information sources. However, information-oriented institutions like the West African Library

Association (WALA) have been set up to contribute to the region's development. WALA was formed in 1953 by neighboring countries—Gambia, Nigeria, Ghana and Sierra Leone—to improve librarianship and information quality among member countries and to foster the dissemination of scholarly information in sub-Saharan Africa. According to Kisiedu (1995), WALA started the *West African Journal* in 1954, to provide information on library-related events, members, and proceedings of WALA conferences. However, they have not been effective, due in part, to political instability in the region. Additionally, there was limited emphasis on information acquisition and limited effort in seeking and disseminating information through the pioneer journal, even in that era when African colonies were approaching independence. The *West African Journal* should have been an agent of nationalism and member countries should have used the journal to spread freedom and patriotic ideas.

Nevertheless, some African academic institutions have become involved in the acquisition, storing, and dissemination of electronic information. The University of Dar es Salaam has since 1994 been producing a library journal which contains literature on CD-ROM technology and micro-CDs-ISIS for information specialists: librarians, documentalists, and researchers. The journal also serves as an outlet for discussion of other IT issues.

New incentives in the library profession might increase continental interest in the information revolution. For example, the Nigerian branch of WALA helped to raise the salary scales of chartered librarians to levels commensurate with the federal professional scale in Nigeria. It also convinced federal authorities to establish a national library of Nigeria. Also, WALA-Ghana received funds from the Carnegie Foundation to train library assistants in the United Kingdom (Kisiedu, 1995, p. 4). In Cameroon, public libraries have been set up in major cities, funded by the government. When more African library authorities join WALA and attend workshops sponsored by the AAAS and other organizations, their efforts would set the pace for Africa's participation in the global information revolution. Since governments have been exploiting indigenous resources to sustain information infrastructures, like the library, before seeking cooperation from abroad, one may expect a major investment in the information economy.

Computer Use

Despite its significant role in bringing information to people, computers might create reliability problems for the new university clientele in Africa. The security, privacy, and legality of sending data across borders remains an important problem. E-mail messages are not transferred through one system, BITNET facilitates automatic searches for the National Library of Medicine (NLM) in Bethesda, Maryland-USA databases. According to Kiathe, (1994), a medical librarian at Nairobi University, the search is prepared by using Medsoftware before it is sent to National Library of Medicine (NLM) via e-mail, after which search results are returned also by e-mail. Users are charged for completing the database search and for services funded by the Pan American Health Organization. Although there is no charge for the use of AIDSline searches, African researchers are at risk of subjecting their work to e-mail bandits and computer "viruses". The fact that BITNET does not have an Internet program makes its users vulnerable. Another Internet client may obtain researchers' findings and use the information to his/her advantage. The fear of burglarizing scientific information during message transmission to or from NLM raises new questions about BITNET's real importance to African researchers. What might happen to a search—a chemical data needed by Researcher A to conclude findings for the cure of a disease like cancer? Although the same search can be made many times, African scientists, unlike their European colleagues, lack the technology to make many searches because electricity supply and cool temperatures necessary for efficient computer operation are scarce. The African scientist may be unable to have legal protection of his/her property, due to the absence of funds for legal matters, international laws protecting the use of computer-mediated information, or an undependable local judicial system. Locally, however, the researcher has the infrastructure to improve research skills. PADISnet, the fidonet is available to experts in Africa for knowledge sharing and discussions. Since PADISnet is linked to the global NGOnet, researchers can hold conferences. That process would give them the opportunity to obtain new information for the designing of appropriate development programs in their various regions, However, the establishment of user privacy and sending data across borders remains an important problem.

A larger problem exists between users, providers, and regulators providers, of the internet. Although as a participatory technology, it is capable of linking and maximizing community spirit, the internet has created a virtual community. People have become more mechanical because they use machines to interact with humans rather than communicate directly. This practice may make people less friendly and less humane. In Western culture, individualism has been maximized through computer access. Africans for whom communalism is the norm, would be obligated to imbibe a new culture and vice versa. Similarly, practitioners in communist cultures will eventually inculcate individualistic ideas through the internet. Because the majority of internet users are Europeans and Americans, with Africans in the minority, the latter's values would quickly evaporate. There is one web site per 500,000 people in underdeveloped countries, as opposed to 10 per 1,000 in America. Those from a stable economic background would benefit more than others.

While American youngsters have adapted to computer use, the educated and financially capable ones in developing countries are only becoming acquainted. Children in American towns have access to computers. They play games, chat with others, develop creative ideas, or surf the internet for information. Children can share the material so gathered with their friends, colleagues, or parents or pass it on to the next generation of computer users. While younger people in industrialized regions are benefiting from it, the older generation in underdeveloped regions that will soon pass away before major innovations in information technology are made, would be able to transfer relatively limited computer-oriented knowledge to the younger ones. Also, because African governments are hypersensitive to public-interest information that they censor it, the allocation of web sites for public use would be difficult. It is also doubtful whether the telephone can become an alternative to the internet for the spreading of development messages nationally. Although a computer terminal is more expensive than a telephone set, African governments could protect their security by adopting the internet and monitoring information transmission rather than striving to control telephone transactions. By leasing or offering free web sites to the public, government secret agents can monitor public opinion regarding its performance. Based on opinion polls, a government can decide the appropriate course of action in order to improve its services to the nation. Conversely, monitoring telephone transactions would only yield limited effects. Apart from tracing calls from point A to point B, secret agents may not have any significant techniques of tracking down public opinion, unless they are investigating a crime. It is easier to read statements on a computer screen than to intercept, or eavesdrop on, telephone conversations for the following reasons: web site users might not have e-mail addresses, so they cannot be traced, whereas telephone users can be identified through caller IDs, or by tracing a phone call. Even though telephone companies in most developing countries do not have caller IDs, and other modern tracking systems, they keep address logs of, and process telephone bills for, resident and business telephone owners. Thus, for developing regions, financial and political power play an important role in the adoption of information technology.

Installing Communication Infrastructure

NGOs and other organizations face geographical, cultural, and financial obstacles when installing telecommunication infrastructures. In some regions, mountains, forests, deep valleys, and rivers make the installation of electric poles or the digging of trenches for fiber-optic wires difficult. On the West African coast where AT&T Corporation has been planting a fiber optic network, inhabitants have reportedly unearthed the wires to design jewelry. Certain villagers have been hostile to engineers under the pretext that the government, which mediated the installation of such appliances, seized their land and destroyed their sacred forests without compensating them. Village residents have halted further installation while other projects have been terminated because of witchcraft practices. Moreover, deforestation destroys the ecosystem, shrines, and other sacred places which these people depend on heavily for their livelihood. Telecommunication installation in rural and semi urban areas obviously threatens, and adulterates, the cultural values of most residents in developing countries. It obstructs the serenity of their ancestral environment and pacifies traditional authority. Without holding meetings with village authorities prior to breaking down the jungle, the governments and companies create distrust among the villagers. That is also why village residents sabotage the installation procedure.

However, things are slowly changing. As a result of coming in contact with that form of Western civilization—telecommunication equipment—and by attending government and church-owned primary schools, interacting with peace corps volunteers, missionaries, nurses, UN peace keeping forces, healthcare centers, the local market place, and taxi drivers (where they meet township people), these village residents have been changing their worldview. They are becoming fascinated with the magic of Western technology and the services it renders. Most local men seek employment in foreign companies based in the village. They take medical examinations and undergo temporary training in order to perform services ranging from the digging of electrical and satellite poles to connecting electric wires. Others have found electricity supply in their villages a privilege. For these people, "civilization" has arrived. In general, the installation of telecommunication equipment and execution of services pertaining thereto upgrade the local economy and reduce rural exodus. This practice could indeed reduce prostitution and the searing crime rate in cities usually caused by idle young men from the villages. The presence of more expatriates in those areas might generate a cultural cross-fertilization, as men and women socialize. However, the high work ethic of this agrarian community might reduce as more people would prefer to spend more money in bars with electricity, than prepare for agropastoral activities. Electricity in the village may also attract crime; hooligans and lazy ones may prey on the fortunes of others in this new marketplace. Despite these problems, local businesses would generate more income and profit. The prospect would be able to purchase basic amenities from the store next door and save time and money, rather than pay a fare to the township to purchase the same products.

Those implications put aside, the entire African continent should be electrified to give the population a chance to utilize new communication technology which is widely available in other regions.

Conclusion and Further Suggestions

The future of computer-mediated communication lies exclusively in the availability of the amount of energy supply. In the 1990s, the supply of hydroelectricity power has been limited. The per capita consumption of energy in sub-Saharan Africa is the lowest in the world, and it continues to grow relatively slower than in other developing regions. While India, Chile, Argentina, Brazil and other developing countries have seen a rapid growth in industry and electricity supply, Africa has suffered a loss due to political crises; wars and frequent changes in government. In 1991, the average person in sub-Saharan Africa consumed about one-third of what an Asian consumed and one-seventh of a Latin American's. Davidson (1994), reported that modern fuel consumption per capita in

South Africa grew by 45 percent , as opposed to 141 percent in Asia. One can attribute this low energy consumption rate to Africa's high population growth rate, high poverty rate, and heavy dependence on charcoal and other biodiversity products for energy supply. Although oil is one of Africa's chief exports, not more than 5 percent of Africa's population can afford its rising cost, or that of electricity. Businesses close down early in the day in order to conserve energy. However, unexploited fossil and other energy resources can indeed satisfy the people's energy needs. Since over 80 percent of electricity produced in Africa comes from fossil, fuels and approximately 65 percent from hydro resources, the continent has the capability of utilizing electricity along with computers to process, spread and share information locally and internationally, if the commodity is properly tapped.

The need for increased productivity and self-reliance arising from exposure to the new world order and, consequently, the interest in emulating capitalist practices, will increase the demand and supply of energy in Africa. Africa cannot participate effectively in a global economy or influence global information technology in the next century if energy supply does not increase by at least 70 percent. No business can be effectively transacted between local, national, and international companies without enough electricity and related energy supplies. The fast-growing population in developing countries, especially rural areas only increases the need for additional energy supply. During the next decade, rural residents will no longer depend on products like wood and coal for energy supply because of massive deforestation and high population growth. Already, the population of the sub-Saharan region is expected to increase by 5.4 percent by 2030 AD—faster than any other region. It will always be important to use new communication systems to expedite the sharing of life and to sustain ideas.

Companies should produce more affordable beepers, cordless phones and fax machines. Although most people may not afford their costs, development information from computers will ultimately circulate among them. Information obtained from the internet designed by organizations and local experts could be disseminated among rural residents at public gatherings. While such information can lose its originality through the two-step flow, information properly processed by NGOs and development communication experts is useful, provided it contains how-to instructions and concepts pertinent to the peoples' values and needs. Authorities should realize that the people in this context live in a world outside world events like sports and terrorism which dominate the dominate the internet. While such information may be fascinating, it is not pertinent to the immediate needs of rural residents--clean water, good farm-to-market roads, more nutritious food, viable fund-raising techniques, more credit unions, better farming methods, and accurate information on wellness and causes and solutions to chronic diseases. These conditions can be improved if specific communication technology is imported and used by developing countries. But which kinds of technology have been used in the pre-electronic information age? What types of technology should be imported, at what cost to the government and to the provider/exporter, and when? Can the technology be sustained? What procurement policies are necessary? Are the citizens fully prepared to adapt?

Those issues are examined the next chapter.

Chapter Five

Exporting Communication Technology to Africa

Overview

Since the beginning of time, exportation has remained the lifeblood of trade. Groups, institutions, corporations and individuals strengthen or weaken their diplomatic, business, economic, cultural and political ambitions through exporting and/or importing ideas and physical resources.

Exportation has significant advantages and disadvantages. Also, if an export product fails to meet the socioeconomic needs of consumers but satisfies the appetite of the exporter, it becomes an imperial agent. In the latter part of the 20th century, groups and nations have been manufacturing and distributing communication technology at an alarmingly fast rate, to maximize profit, intensify competition, and enhance or maintain control of other entities. Following the deregulation of the telecommunication industry in most Third World regions there will be increasing competition in the provision of services. According to a 1996 study conducted by EHE InfoLink, a telecommunication corporation based in North Potomac, Maryland, over 700 million dollars in telecommunication revenues annually will be open to competition among providers seeking Third World markets. That means more countries and corporations will be importing and exporting services and equipment. Therefore, the need to investigate the communication products being exported and the exporter's bargaining power cannot be ignored.

In order to evaluate an exporter's bargaining power, we must identify from a list of other factors the technological mindset, which produces the product. Knowledge applied in a systematic way to achieve control over nature and human patterns of behavior, is technology. But the term "technology" does not refer only to goods or production facilities but also to the ideas needed to manufacture them. That means an idea is needed to create and use a product. The idea determines how the product should be used and by whom, when and why. Ideas are created through observation and imagination. An idea is conceptualized and tested before a product is invented. Human society is no stranger tothis systematic development of knowledge. In ancient times, land was the primary resource for generating culture, new ideas, and wealth. Before the Industrial Revolution, such investment took precedence over anything else. Today technology and information science are the major resources for generating efficiency in satisfying human needs. Understanding the importance of these innovations require an indepth study of sociocultural trends (Ngwainmbi, 1997, 1998, 1999) and a chronological analysis of ideas and resources. Root (1968) argued that the promoters of new systems for diffusion rely on the following assumptions: (1) an innovation can only exist after it has been diffused (2) the diffusion path can only be traced from the initiator to the adopter, and (3) the diffusion is directed for training and dissemination. The third assumption is based on a larger concept: training and dissemination can only take place after the technology has been imported and adopted. The patents of the importer, the sale and use of the product, are all determined by the quality of the product, exportation cost, need and purchasing power of the importer. These paradigms generate ideas and information for the importing and exporting entity and facilitate production and marketing.

Innovations should be adapted and diffused by local entities to meet changing needs of their consumers. However, access to such resources as scientific knowledge is not the only prerequisite for the creation of new technology. Indigenous technology can be created, developed, and diffused to meet basic community needs. It takes more than capital, access to scientific knowledge and research centers to create a new technology. But the creation of a new technology is not a farce: It demands cooperation among government and industries, links between the educational and productive sectors, a sound financial base, and the creative energy of a country's population (Singer, 1977). Local residents and interest groups must be prepared to share and apply their knowledge of a technology to curb the influx of products.

Meanwhile, the Western Hemisphere remains the primary exporter of telecommunication technology to Africa, from the colonial to the neo colonial era. The first exporters of telecommunication technology were the missionaries. This group sent radio sets to fellow evangelists in Africa to listen to "home" affairs and religious information. The short-wave transistor radio sets with strong frequency modulation transmitters enabled the missionaries to listen to B.B.C., V.O.A, and German, Portuguese, Spanish, French, Italian and other foreign programs. Also, through that medium, they received and shared information about missionary activities in other local regions. Some administrators used sartorial communication to supervise the work of Africans in the fields. For example, in *Bafut Beagles*, Lawrence Durrell describes a scene where a colonial administrator left his sunglasses on the farm and went home, but the farmers continued working whenever they saw their reflection in the sunglass. They thought he was still watching them.

Local Information Technology

Tribes coexisted and managed military, social, and economic affairs using hand-made technology like drums, flutes, iron bells, and guns. A powerful telecommunication hardware, the "talking drum" comprising wood andanimal hide has, for centuries, been used in sending and receiving messages from long distances in rural areas. Nigerian researchers, Arewa and Adekola (1988) have said that this gadget can evolve into a modern theoretical paradigm. Adekola (1989) predicted that in eleven years, Nigeria (and other African countries) will reach a stage where it (they) can use locally- produced spare parts and equipment to sustain a digital and analogue telecommunication system (p. 20). Since it is already a popular means of communication in Africa, the talking drum could become one of the most viable technologies invented by Africans for rural African residents. It was used to alert village residents of imminent military invasion, a tribal war, a village festival, an obsequy, or a court summons. The sound of the drum clearly indicated a specific message and the group being targeted understood the message and took necessary action. Drum manufacturers should think of innovative ways to process the animal skin, wood, and cords in order to produce louder sounds for distances beyond at least a ten mile radius—the farthest distance at which drums are usually heard.

Gunshots, flutes, and other sound-making products were used by both natives and colonial administrators to inform people of a death, a festival or any cultural activity requiring the participation of all village residents (Ngwainmbi, 1995b). Members of sacred societies also used flutes to advise people living apart of meetings and other group activities. The notes of the flute had a specific message, which could be decoded only by members of that society. Among the Bali Nyonga, who are parts of the Tikari tribe in Cameroon, members of *Voma*, an all-male sacred society, drone through the community once a year to cleanse it. The *Voma* swing a special sling that growls and roars throughout the village. Women are not allowed to see it or imitate its sounds, as heavy penalties, including blindness, might befall them. In most grassland regions of Africa, hunting dogs had jingling bells attached to their necks to prevent other hunters from mistaking the dogs for animals of prey. Such bells were also worn on children's ankles to keep parents informed of their location.

For centuries, bells and heavy metals have been used for instructional purposes. When the missionaries and administrators erected adult learning centers where they taught arithmetic and writing, they used jingling bells to identify time or to signal a change in a school activity. On a regular school day, the bell would ring approximately six times: the first bell would invite pupils to school, the next one would send them to class, or inform them to step out for a brief recess, lunch, manual labor and go home.

Blacksmiths invented long, thick, heavy metals that supplemented the bell. This communication technology better known by the onomatopoeic term "gong", constituted a long metallic rod hooked to a tall tree. A designated timekeeper would strike it several times using a short metal rod. The number of strikes expressed a specific message, which the pupils understood. A series of evenly paced strikes, usually at 7 o'clock a.m., was a wake-up call for them; the second series at 7.30 a.m. were more rapid, and they reminded them to come to school. The third, most rapid series advised them to rush to the assembly ground for prayers and announcements; and the last strikes signaled closing time. Between school hours, which lasted between four and eight hours per day, a gong indicated a special event. The gong could reach pupils in four to five villages, approximately 10 miles apart. That communication technology, reminiscent of the alarm system, is still in use in some rural primary schools in Africa. Mostly elderly and skilled people used the flute, (a 12-inch oblong object with six holes designed from Indian bamboo or brass, as a musical instrument). It was manufactured in large quantities and made available to everyone, and used in the communication of entertainment messages or in the performance of other social functions. Although the telephone remains the fastest and most effective channel for transmitting interpersonal messages to and from long distances, the flute would have been able to serve a smaller human community living together, if blacksmiths still had the incentive to design sophisticated ones. Since there is no competition, simple operation methods and a relatively small market, the flute has become a less demanded and less marketable commodity than the telephone which is in high demand, is used by many people in the world, and is available in any electronic marketplace. Even in poor rural areas, the economic value of the flute will remain low compared to other state-of-the-art sound equipment: deck, cassette recorder, cassette, and VCR—manufactured in technologically advanced countries that control the stock market.

The use of these and other indigenous technologies diminished because the colonialists prevented indigenous products from competing with European products in any market. In 1905 before British manufacturers started supplying their products to their colonies, they set an agenda that would replace the marketing of such locally manufactured products as spades, forks, shovels, steel files, hammers, axes, hoes and hunting bells. The importation and heavy use of alternative metal through the collaborative efforts of high level officials in the colonial government crippled the incentive of local blacksmiths who had the ability to forge state-of-the-art knives, shields, tin, flutes, swords, and armors, among other hardware. These metallurgists whom the aristocracy, farmer, and king's court depended on for war weaponry, farm tools, and media equipment , respectively lost a substantial amount of their income during the importation, processing and marketing of scrap iron. It also affected the way war was being waged and diminished the need for manufacturing weapons.

Even the jingling bells that are used for hunting, to mark time and identify children in Third World rural societies are not widely used in societies with large economies and with culturally diverse people, except in some traditional Catholic churches or rural settings. Conversely, telecommunication companies in industrialized societies do not find a need to develop the bell or flute into a sophisticated technology, because it does not have any significant market value. Instead brass bells have been replaced by car- and house alarm systems, beepers, fire-quenching trucks and police sirens, which are more portable, more audible and certainly more in demand in a fast growing industrial and populous world.

Bells and flutes have proven effective in transmitting mostly entertainment messages among traditional Africans, and they are less costly, easily available and are still being used in most regions in Africa. Because people are familiar with their functions, the flute and bell technology will eventually give way to post-modern technology, for the following reasons:

1. In contrast to post-modern telecommunication technology, only a select few operated those media or know how they are used. But because more Africans are becoming interested in foreign culture and foreign goods, they will minimize the potential of local technology. Also, more people believe that foreign goods are more durable and more attractive. If someone else wants them to use their product, it means that person is confident of his/her product.
2. More people would be able to use new communication technology thereby expediting the communication process. This would enhance the creation, dissemination, and exchange of innovative ideas and information. It will also reduce ignorance and augment knowledge of personal hygiene, creativity and interpersonal interaction and will eventually enhance the individual's economic status.

Under colonial administration, blacksmiths, the backbone of the economy who formed the majority of the local middle class, were forced to pay more taxes, which diminished their incentive to develop marketable technology for mass consumption. Taxes for such first-class occupations as black and white smiths were doubled because they were considered first class citizens, leaving the producers with very little funds to survive. As there is no comprehensive record of how the tax money was invested, one can only speculate that the colonialists spent it on their families or shipped it to their home governments, instead of using it to provide services needed by local residents. Local technology was, thus, through high taxation and stringent importation policies forced to succumb to colonial capitalism. This practice lasted over 100 years in some African countries, which partially explains the availability of underdeveloped and less marketable local telecommunication technology like flutes, bells, guns, and drums.

The governments of African countries seem to have realized the dire need to remove their nations from a state of economic bankruptcy by allowing foreign telecommunication industries to invest in Africa. But just how committed are they to making sure Africa is included in the global information marketplace? This question could be partially answered by analyzing foreign competition.

INTELSAT and Its Services

Arthur C. Clarke generated the concept of inventing a land-based system that would have a global impact in 1945. In his article published in the *Wireless World* titled Extra-Terrestrial Relays", the scientist theorized that three satellite systems in geosynchronous orbit could provide coverage to the world. It took thirteen years to put it into practice as the first active communication satellite—SCORE—was launched. SCORE received a message from earth, stored it on tape and later transmitted it. In 1960, the first live two-way voice communication took place via satellite. Two years later, the first satellite, Telstar was able to use a travelling wave tube power output amplifier, which demonstrated the communication techniques and equipment to be used in commercial systems, and led to the invention of a live television transmission. In 1963, Syncom II, a satellite system demonstrated the ability for ground stations to provide continuous coverage. This showed that a medium altitude commercial system could be established. In August 1964, eleven nations signed an agreement to start a global communication system—INTELSAT. This milestone was later acknowledged by America's President Lyndon Johnson, who stressed the potential for the technology to improve communications between people and nations and to promote commercial service between continents.

INTELSAT has become one of the most powerful satellite service providers in Africa. With an ability to provide connectivity to over 200 partners around the world and a multiple in-orbit fleet resource comprising 25 satellites, INTELSAT could enhance Africa's access to the information society. With the deployment of its 805 access to 33 degrees E of the orbited location, it covers the entire African continent. In 1996, INTELSAT evaluated the deployment of future resources on an ongoing basis to ensure that its capacity was positioned where it was most needed (*INTELSAT Annual Report*, 1996, p. 5). Since economists evaluate product success based on customer demand and contingency planning, and existing customers receive higher priority in service than new customers, it is puzzling that INTELSAT considered Africa a fast-growing market for its satellites when most Africans do not have basic telecommunication electronic equipment. Although the world's largest satellite system is located in Central Africa none of the countries in that region have any comprehensive telecommunication system, or control over its operation policies and activities. The deployment of its services can only help foreign scientific and mass media institutions in capturing highly marketable images. Additionally, INTELSAT can be used for the advancement of other technologies.

INTELSAT was, however, enthusiastic about increasing its services in Africa between 1970 and 1990. It provided interconnectivity services to 49 nations operating through 210 earth stations. That reduced the need for transit traffic outside the continent. INTELSAT also accomplished the following: networked with the Central African Customs and Economic Union to improve interconnectivity among member nations, identified ways of improving interconnectivity by satellite links in transit centers of each region, and assisted countries in developing small aperture terminal (VSAT) in rural areas, to complement their domestic networks. Its satellite systems were increased from 332.5 degrees East to 335.5 degrees East in order to fulfill requirements for greater connectivity to other regional areas. Transit centers exist in the four main regions of the continent. Generally however, INTELSAT has not provided any significant service in Africa between 1990 and 1998. This partially explains why international network systems in most of the countries are still underdeveloped and far less efficient compared to other Third World areas like Singapore and Brazil. One expects a stronger commitment by

INTELSAT to improve the rural and urban sectors of the continent, if African countries make significant investments in the management of the earth station, and when more foreign business institutions continue to demonstrate a strong interest toward African telecommunication management or make heavy investments in Africa.

Foreign Competition

According to the U.S. Chamber of Commerce, the U.S. exported over 5 million dollars worth of telecommunication equipment to 51 African countries between 1990 and 1994. We may easily be impressed by such efforts made by foreign companies toward the exportation of telecommunication export products to Africa. However, we must understand that most of these companies only recently began considering Africa a big market through unusual circumstances; the release from prison of Nelson Mandela the world's most famous prisoner from South Africa, lobbying efforts of religious and humanitarian groups around the world, and death-based democratic activities in African countries. The historic release of Mandela from prison and his election as South Africa's first post-apartheid president along with some political stability in that region brought renewed economic activities from world companies. Today, communication technology ranks as one of the highest imports in South Africa since 1990.

A breakdown of telecommunication equipment shipment to Africa shows Japan as the leader because the Japanese have a reputation for manufacturing and exporting low cost but durable products to Africa. Automobiles, corn grinding, forestry and sewing machines are shipped regularly with affordable parts. Not only are 75 percent of automobiles used in Africa imported from Japan, but also so are television sets, radio broadcast receivers, and sound recorders. Most homes use Japanese-made "Toshiba" and "SONY", and German-made "SAMSUNG" television set. According to a 1994 U.S. OECD Foreign Trade Statistics, France has been the second largest exporter of sound recorders to Africa, followed by the UK and the US. The report also states that the U.S. is second to Germany in the exportation of general telecommunication equipment. In 1985, the U.S. sold $80 million worth of telecommunication equipment to Africa, but that dropped in 1988 to less than $40 million. In addition, the U.S. has played a leading role in the exportation of telecommunication technology to sub-Saharan Africa, its exports there were valued at $69,365,000 in 1990. During that year, television, radio and sound recorder receiver parts totaled $21,719,000, followed by transmission and reception equipment.

After the Gulf War in 1991 until 1996, U.S. telecommunication companies were among the foremost exporters of telecommunication services and products to Africa. Some U.S. companies have formed partnerships with African telecommunication entities in order to facilitate the transaction of telephone services. But the telecommunication market in Africa has been inconsistent. Most African consumers studied for this book assert that they use the services provided by American long-distance companies because the latter offer lower rates than French and German companies. However, South Africa has remained one of the largest markets for U.S. telecommunication products, with eight of the top ten telecommunication exports since 1990 originating in the US. Between 1990 and 1994, South Africa spent over $5 million and it remains the largest market for importing radiotelephonic reception equipment, for carrier-current line systems, telephonic switching equipment, telephone sets and telephone equipment (OECD, U.S. Chamber of Commerce, 1994, p. 4). According to the same source, Nigeria, which spent 9.5 million dollars on transmission and reception apparatus from the US, was also the largest importer while South Africa and Zaire spent over $1 million on the same equipment. Burkina Faso, which accounted for 25.6 percent of Africa's market, was the largest importer of U.S. receiver parts, spending $5.5 million. Lesotho, Africa's smallest country in size and population, spent about $1 million in receiver parts to upgrade its earth station. This expenditure is rather small compared to other products imported by other African countries, which show that African leaders do not realize the importance of prioritizing telecommunication development over other needs to bring their countries to a competitive level with other countries in the world. African leaders are being criticized especially by the elite (educated class) for not using their authority to import and upgrade telecommunication technology But they do not seem to understand that certain procurement policies hinder the exportation of local products.

Exporter's Procurement Policies

It is not easy to ascertain the nature of exporters' policies as rates, markets and consumers change constantly depending on the competition and consumers' purchasing interests. Therefore, this author will not pretend that the policies to be discussed herein are appropriate for all exporting agencies because public disclosure of certain policies could diminish exporter's profit or raise suspicion among consumers, potential markets and competitors. Nevertheless, there are critical factors to be considered when explaining an exporter's procurement policies. The biggest impediments to the establishment of large markets are budgetary constraints and fluctuating foreign currency exchange rates. In Cameroon, between 1995 and 1998 when the economic crises was at its peak, the exchange rate for $1US was 500 francs, CFA, as opposed to 275 francs, CFA between 1991 and 1994. This constraint as well as transportation costs for telecommunication equipment have influenced decisions about the importation and use of telecommunication equipment. Although purchasing and transportation assistance may be provided by multilateral institutions like UNDP, IMF, and the World Bank in development assistance, telecommunication importers cannot continue to depend on donor institutions to support them. Moreover, some donors implement tough policies after agreeing to assist the country requesting their help. While some experts do not see any foul play when an exporter influences the policies of the importing country, it is important to avoid policies that hamper the smooth and sustainable operation of telecommunication systems in developing countries. And to avoid them, countries importing such equipment should consult foreign commercial banks for import financing with clear cut plans to pay off loans and avoid incurring overly burdensome interest debts.

U.S. Telecommunication Export Policies

U.S. companies have been handling business relations with most of their Third World partners through intermediaries, one of them the Matchmakers Trade Delegation (MTD). This organization matches U.S. companies with potential agents, distributors and licensing partners. In addition to helping small and medium-sized companies meet their export sales objectives efficiently and economically, registered as a private corporation with the U.S. Department of Commerce, evaluates the potential of the exportable product, locates and screens contacts, provides in-depth briefings on the economic and business climate of the country being visited, and organizes trips and face-to-face meetings with prospective clients. According to a 1995 news bulletin of the International Trade Administration, MTD also provide the following services:

1. Prescreened prospects interested in exporter's product;
2. In-country publicity;
3. Convenient sales avenues;
4. Business appointments scheduled for exporters through U.S. embassies or consulates;
5. Thorough briefings on market requirements and business practices; and
6. Interpreter services.

The company's 1995 schedule included projects like the transfer of environmental healthcare technologies to Brazil, Chile, Argentina, Singapore and Thailand.

The U.S. Chamber of Commerce, in its efforts to emerge as Africa's most significant business partner, in collaboration with MTD outlined policies for participating in the matchmaker program. These policies do not only ensure that U.S. exporting companies protect and maintain U.S. international relations, but they also monitor healthy and profit-making operations for the U.S. companies. An outsider may argue that the policies benefit the importing agency because it has the capability of promoting its own licensing agreement and of determining agent sales or distributor policies prior to engaging in any transactions with the exporter. Clearly, the policies do not include the importer's expectations; hence the importer operates at its own risk.

Exportation Problems and Prescriptions for Solutions

Financing Telecommunication

The policies of some governments virtually prohibit the use of government-owned telecommunication equipment for commercial purposes. For these governments, allowing telecommunication entities to operate as commercial enterprises is like surrendering power to the citizens. Government policies on telecommunication operations are insensitive to the very rationale for their operation. Because government has complete autonomy over imports, the telecommunication ministries do not yield sufficient profit. There is no incentive to contain cost or improve customer service, neither does it create a forum for competition, quality service or career opportunities, since government workers treat the workplace and equipment as their personal property.

The deteriorating condition of telecommunication equipment in Africa calls for major changes. Apart from using antiquated and damaged equipment imported mainly from France decades ago, the personnel are undertrained and the work ethic is poor. Not only do workers come late to work, they also leave early. Some workers are bribed to use private telephone lines in making overseas calls for themselves and others. Residential phone lines are issued not to garner profit for the government but to appease friends and family members. Also poor maintenance of equipment, slow action toward installing new residential and business phones, slow response to changing demands in work ethic and technology advancements account for the degenerating state of the industry in Africa. It sometimes takes two years install a new residential phone line, even after the government worker has received a bribe from the customer.

These are the reasons behind the decision by African leaders to buy services from AT&T, Afronet, ALCATEL, and other foreign telecommunication agencies. Through negotiations between African Ministers of telecommunication and AT&T executives, the latter embarked on a 20-year trial fiber-optic project with the hope that local governments and other foreign companies would form a partnership for investment. AT&T's fiber-optic loop, which costs approximately $1.5 billion U.S. dollars may include an 18,000-mile undersea cable encircling Africa (Burgess, 1995). Besides bringing the possibility of increasing economic growth, the completed project would have the capacity to carry about 90,000 conversations simultaneously. Certainly, the undersea connection will increase the frequency and volume of telephone connections and fax transactions and Africa's chances in the global marketplace.

But how will this project whose connection cost was estimated at $2 billion be financed? The World Bank, multinational corporations, African countries and international long distance carriers like MCI and Sprint have been co-financing it. However, foreign investment in the industry has been very low because of foreign currency shortages in the public sector and weak domestic currencies. Most African governments prioritize such issues as public health, transportation, sports, and entertainment over telecommunication development because they do not recognize its importance in facilitating international trade and international communication, and the necessity to meet basic human need.

The internal organization of telecommunication enterprises in Africa follows government, not business principles. This causes severe weaknesses in organizational structure, financial management, accounting and information systems, procurement practices, and human resource development. Although they have been privatized, the government is still supervising the telecommunication ministries of Cameroon, Senegal, Ivory Coast, Chad, Central African Republic, and Zaire.

To create a forum for the financing and sustenance of the industry, African governments must liberalize the telecommunication sector. They must raise the telephone call completion rate from below 20 percent to about 75 percent , by allowing major private phone companies to install and operate their equipment for several years. To meet world standards by the year 2000, African governments should invest at least $30 billion per year on basic telephone service. The introduction of competition among long distance telephone networks would increase traffic volumes, improve the quality of service, and reduce rates for subscribers. In addition to liberalizing the industry the governments need to run a public network that would serve college students, public schools, and poorer citizens. Licensed private companies should be connected to the public network in order to give "public users" access to their products and services. If an independent agency monitors their activities, both private and government provider would ensure quality and low-cost services to the customer. However, reducing cost may not guarantee any successful long-term business. Through strategic and financial investment, both providers can increase the market because too much government control has always discouraged investors and too little government has caused financial and management difficulties. Strategic investors should be given a large percentage of equity in order to ensure a sustainable interest. Although this may mean a long-term commitment from the government, to ensure the manifestation of the exclusive rights of operation for investors, bad government policies and the ever-recurring political upheavals may cause investors to withdraw.

African governments do not address efficiency, growth, or immediate fiscal gains, rather their policies seek to consolidate government's economic and political control over the citizens. Political unrest in Africa has substantially reduced foreign investment potential on the continent. For example, in Cameroon in 1992, when presidential elections were rigged and public demonstrations and civilian massacres ensued, foreign investors stopped the implementation of a free trade zone in that country. Boston Bank and other major foreign companies closed down abruptly, leaving the business community in economic ruins. Certainly, any financial investor would not want to lose funds in an unstable environment. However, policymakers can rescue such situations by advising local and external investors on the stability of the country's currency. In so doing, they would have created an appropriate environment for the financing of telecommunication projects. Lomax (1995) advocates "a sound and growing economy which is not overly based on a single commodity; stable currency, a declared policy base and established objectives for the telecommunication sector (The objective should be growth and efficiency not immediate fiscal gains); clear commercial, telecommunications, investment code, legal frameworks; a transparent regulatory framework (and), clear commitment to sector reform from government and sector entity incorporating market structure, ownership and options, pricing, etc." (p. 15). Based on those conditions, countries that are financially indebted to the IMF and other world lending institutions would become eligible for financial assistance. However, African countries will continue to be dependent on international lending institutions until African governments fortify their own currency and economy. With the exception of Ghana, no African country in the Sub-Saharan region has had a stable currency or a growing economy in the 1990s. Nigeria, which had one of the strongest currencies in Africa in 1993, has since had one of the weakest currencies in any market despite its vast economic resources. Based on Bank of America records of January 8, 1993, the exchange rate for one U.S. dollar was 21.70 naira. That rate has only tripled since 1993. The power of the CFA francs (used in French-speaking African countries) has dropped from 279 francs in 1993 to 600 francs CFA per $1 between 1996 and 1998. Although the

IMF and other lending bodies have lifted its borrowing sanctions on some African countries, provided loans, and sponsored major projects locally, its ethnic warfare, unstable government, mass corruption, limited business relationship with foreign companies, fluid policies, and a weaker currencies have all reduced foreign the investment rate. Part of the problem may have been caused by "gray", stringent conditions imposed on them by the lenders. For example, most World Bank financing policies are lacking in depth and tend to encourage more indebtedness for the lender than assist the latter in establishing economic independence. By providing large loans and high interest rates, it becomes difficult for borrowing countries with low GNPs and weak economies to pay off their loans and interests within the stipulated time. Additionally, the borrowing countries are often asked to fulfill other requirements that eventually cause more internal economic and political crises. In the early 1990s, for example, the Bank wanted the Cameroon government to reduce the number of government employees—civil servants—to one third to be eligible for any further loans. Although downsizing would eventually enable governments to invest civil servants' salaries in other capacity-building ventures, it created major financial and social crises for most citizens, in the following ways:

- About 80 percent of the workforce constituted government employees and much of the population was dependent on the latter for economic survival. Put succinctly, government employees supported 70 percent of the population.
- Given that Cameroon is a consumer-oriented rather than producer-oriented society, where civil servants rely solely on government salaries to feed, cloth, or sponsor their relatives in educational and vocational institutions, the downsizing of the public sector with no clear short-term alternative plan for catering to the immediate economic needs of such dependents, or stabilizing the local economy only maximizes poverty, misery, and anarchy.

Such conditions have made most African countries eternal debtors to international financial lenders. This World Bank mentality is mainly responsible for underdevelopment in the Third World because of its difficult policies and because it supplies needed funds but fails to sustain the projects for which it provides the funds. Bank officials have

little attention to how its funds are used by the borrower. The only meaningful way of financing and sustaining telecommunication operation in Africa is through investment by private industries like AT&T, MCI, Sprint, NTC. These industries should invest their own money and design most of the operation policies, since African governments do not provide the capital to buy their entire equipment and services. Moreover, telecommunication ministries do not have clear commercial investment codes because they practice corruption in all ranks of the Post and Telecommunication ministries, including employing unqualified friends, cronies, and family members in high level decision-making positions. Therefore, it is difficult to change from a government-managed economy to a free market economy, foster commercial telecommunication activities, or implement legal and investment codes.

Telecommunication development in Africa should be conceptualized and implemented as a business, not a political venture. Politically inclined institutions like the World Bank, the UN, and the IMF should encourage private corporations to carry out market research and plan business strategies with informed people, shareholders and major business managers in Africa, otherwise the information superhighway will bypass Africa as have other modern technologies.

Telecommunication Services in Africa

Telecommunication services in have been beset by obsolete equipment, incompetent personnel, and limited funding. Although the Cameroon government invested about 3.6 billion of the francs, CFA between 1972 and 1982 on INTELCAM, it failed to meet the increasing needs of the modern Cameroonian society. Before obtaining independence from European administrators, African countries had been using telecommunication equipment. In 1936, the elite in Cameroon were able to communicate abroad through submarine telegraphic cable. In 1952, the first telephone communication took place. According to the 10th Anniversary Report of the Ministry of Post and Telecommunication in Cameroon, the French Cable and Radio Service took over the telecommunication industry in 1960. But the Europeans operated most of the equipment. As such, Cameroonians were unable to manage it effectively when they became independent. Even those trained abroad after independence abused the equipment by employing workers with minimal knowledge of its operation. The problems facing Africa are demanding. There are about 151,000 villages, and over 122,000 of this number have no telephone services or electricity. This represents 80 percent of the continent's population most of which is rural. In order to reach the modest goal of one telephone line per 150 people by 2000 AD, it will be necessary to increase the lines by at least 4 million lines, which according to Westendoerpf and Odeh (1990) requires an investment estimated at $6 billion. Another setback suffered by many African countries is poor management of telecommunication facilities resulting from too much government; more administrators than technicians. This makes the technicians (most of whom are undertrained) to become overburdened with repair problems from subscribers.

Additionally, the poor condition of turntables and the scarcity of tape recorders prevented reporters from completing their assignments on time and/or preserving information for further research. Journalists on CRTV (Cameroon Radio and Television Service) like other African communication practitioners have been using the same audio and videocassette recorders purchased before the country's referendum in 1972, to record interviews. Due to insufficient funds, telecommunication workers have not been able to cope with the ever-changing innovations in the industry; they have rarely attended important international telecommunication conferences. Even at conferences abroad, they use outdated equipment like slides and projectors, instead of using multimedia and other distance learning equipment which participants are conversant with. Also, because most of them have not received further training, they are not as familiar with new state-of-the-art equipment as have other participants, their chances of attracting potential foreign investors are reduced drastically.

However, certain governments have been making efforts toward the improvement of the industry. In 1975, the Cameroon government created INTELCAM-*Societe des Telecommunications Internationales du Cameroun*— to offer "a broad range of good quality services that include telephone, telex, telegraph, special circuits, radio transmission and television" according to then Post and Telecommunication Minister Ibrahim Mbombo Njoya. In a 1997 New Year address to the nation, Cameroon's president promised to bring Internet services to the people. Some progress has been made in the industry. After the creation of INTELNET, a micro terminal linked to INTELSAT satellites and the Earth Station, administrators, enterprises and professional organizations have had access to information. It is linked to computers in the U.S. and European countries. In addition, it provides sales information and services for the distribution and collection of economic data on press agencies, printing houses, and business organizations. INTELCAM has been operating an international telecommunication program that includes telephone connections, telex and telegraph transmissions and facsimile. Also accessible to international users for information retrieval and cyber-mediated dialogue is INTELCAM website. With the installation of public telephone booths in Yaounde, Douala, and other major cities, callers are guaranteed smooth connections and operator assistance. The industry also anticipates using circuits with higher connection speeds from 48 kilobytes to two megabits per second by 2000AD and the call completion rate has been 95 percent. Directors of banks, petroleum companies, industrial and commercial establishments in Cameroon and abroad have been using INTELCAM services for the transmission of huge volumes of information and data, however, small scale businesses, middle and lower class citizens who make up the majority of the country may not benefit extensively from INTELCAM services in the near future because of the high cost of international calls.

Unlike PANAFTEL, INTELSAT, and RASCOM, INTELCAM one of six earth stations in the world and the only in Africa sought to expand intercontinental satellite telecommunication by providing telemetering, tracking and command and monitoring satellites to INTELSAT. Despite its inception in 1975, INTELCAM has failed to provide quality services within Cameroon. Data transmission has been slow. Since its installation through 1998, the completion rates for calls made abroad have been 15 percent , and 35 percent for calls made directly from the telephone exchange centers. It often takes eight or more attempts to complete a phone call between towns in the English-speaking region and at about five attempts for international callers, because of an antiquated telephone system installed by the Germans during the colonial era. Phone calls from Buea (a historic town in the English-speaking region) to other national cities have passed through Douala. With only two international lines originating from Buea, the call completion rate for outgoing and incoming calls is about three percent. The lines are jammed between 7 a.m. and 10 p.m. U.S. time. Conversely, call completion rates in the French-speaking region are relatively higher because they have been managed by ALCATEL, a French telecommunication company. If AT&T

completes its installation of the fiber optic system, the call completion rate in the country should improve. French-speaking countries have too many "important" officials in a given organization. INTELCAM, for instance, has over fifteen divisions, most of which perform similar functions. This makes the administration of public policies cumbersome. Moreover, each division has a "chief" and an assistant "chief" who have limited knowledge about the operation of telecommunication equipment and services, but make policies that oversee the operation of the entire industry.

In a country of 14 million people with one phone per 200 people, the need for a powerful fiber optic network cannot be underestimated, especially in the 1990s when Cameroonians like other Africans are pursuing democracy and ways of exercising their rights to free expression. Initially, the government may have been slow in extending telecommunication services to all citizens for fear of compromising its authority. Because Cameroon was one of the first countries in Africa to establish rapid interaction with Ivory Coast, Congo, and other West African areas through a sub-marine telegraphic cable system installed in 1936, one cannot underestimate the government's desire to promote the use of communication technology on a national level. Since 1997, under the auspices of its president, the government sought to accelerate Cameroon's participation in the global information supermarket by extending internet access to non-technology owners. With only 15,000 computers and 60,000 telephone lines in the entire country, (Counting was done in January, 1998) non-telephone and non-computer owners have been keeping in touch with the world through a new program sponsored by the government and philanthropists. From a university-wide computer network, students have free access to research data via world internet.

Since 1989, some French-speaking African countries; Senegal, Mali, Burkina, and Benin, have been making significant progress in order to benefit from the world information supermarket. Under PANAFTEL, a telecommunication infrastructure was implemented to consolidate human resources. The Canadian government disbursed 14.5 million Canadian dollars in 1989 for the financial and technical autonomy of telecommunication organisms in the five countries. Also in 1992, the Canadian International Development Agency (CIDA) donated 852,575 million Canadian dollars to facilitate telecommunication links in rural areas in Zimbabwe. CIDA has been one of the foremost non-aligned sponsors of telecommunication projects in developing countries. Between 1993 and 1996, CIDA disbursed approximately 80 million Canadian dollars for Pan African Telecommunication, Zimbabwe, and Egypt to upgrade their telecommunication systems. The funds were to be used specifically on pilot projects and other institutional necessities.

Countries throughout Africa have been emphasizing the expansion of urban-rural communication units, seeking to increase the number of telephone units. In 1988, there were 6.353 million units; 83 percent of which were located in urban areas inhabited by only 70 percent of the population.

Summary and Further Suggestions

The exportation of communication technology to developing countries has implications ranging from the exporter's *raison d'etre* to the importer's readiness. Similarly, the use of information technology there depends heavily on the kind of product that is imported, its economic value to the local government and the nature of the agreement, or an understanding between the provider and the local administration. However, due to the inconsistencies in financing problems faced by local markets, as well as exportation procurement policies, clear-cut long-term advantages of applying foreign telecommunication products in African countries are not totally feasible. In addition to supporting foreign investment in the local industry, local officials should complement the usefulness of local telecommunication systems, by investing in their markets. Local investors and telecommunication officials should select, deregulate, and promote products like the facsimile and cellular phones, that can provide low cost and highly efficient services to more low income people over a longer period. Although the costs of telephone calls have dropped for U.S.-based subscribers (where a phone call to Africa on an AT&T calling plan costs approximately $1.50 per minute), callers would save money by condensing information and faxing it to their parties. Digital consumers who transact business in Africa should obtain the Worldwide Digital Fax network, which provides transmission with encryption for complete privacy. It would cost fax users less than $1 a page to send messages to or from Africa. Even fax users can bypass telephone lines by using digital systems to send and receive messages. Language should not become an impediment because the sender and receiver must have had some form of prior contact. Businesses in America and Japan, which carry the heaviest volume of fax transactions in the world, have resolved this problem by using translators for faxed messages.

The next chapter expands the scope of understanding Africa's readiness for the exportation and use of communication technology, by examining internal regulatory bodies and further describes conditions for improving telecommunication management in developing countries.

Chapter Six

Telecommunication Policies in Africa

Overview

This chapter examines the changes in telecommunication policies as well as certain characteristics of recent legal practices in Central American countries. It further provides brief analyses of legislative situations and in other developing regions and their relationship to the economy, and offers policy considerations necessary for the establishment of national information networks.

The rapid pace of technological change has had a strong impact on telecommunications activities and policies. Just as cable, television and video cassette recordings changed the role of broadcasting in the 1980s in the Free World, so will newer telecommunication technologies change the way information is perceived or used elsewhere.

Change in the use of communication technology is determined by the policies governing its operation. Since policies are designed based on society's values, their implementation should not be based solely on extraneous variables like a country's economic or military ability, but should be observed by everyone irrespective of race, class, or ethnic background. But what are policies and why must they be implemented?

Policy or "regulation" is the process whereby a governing body formulates rules to mold conduct. Barrym (1980) saw it as a process consisting of the intentional restriction of a subject's choice of activity by an entity not directly party to or involved in that activity. There is a problem with that conceptual definition of this term and serious implications on how it can be applied. The idea that regulation should refer only to the actions of governmental bodies (known elsewhere as regulatory agencies) is biased; it fails to include the actions of civilian groups who constitute the backbone of society. If regulations are designed to oversee the behavior of society or mold private, economic conduct, they must have society's input. Another problem is, when regulation restricts a "subject's choice of activity", it also restricts that of the political institution which consolidates executive, judicial, and legislative functions into a single system. In democratic societies like the United States, regulations are formed by new institutions of power based on the common laws of the former institutions while in Africa, they are based mainly on personal views or informal modes of rule-making. The advent of a regulatory agency corresponds with the rise of a nation's economy and the need for the government in power, to establish effective social measures for progress. The ambivalence in the operational definition of the term "regulation" makes the conceptual framework even more troubling. Hence, any theory on regulation must situate the regulatory agency in a historical or cultural context from which the agency was born.

Policies are usually designed based on culture, defined herein as thought that stimulates group behavior, e.g., needs and values. No institution designs policies, which it cannot implement, however, the implementation of all policies has political and legal underpinnings. In countries with pluralistic governments, judicial, executive, state, federal, and local bodies regulate broadcasting. In the US, Congress designs most telecommunication policies, which are enforced by the FBI, FCC, FTC, and other state and local agents and interpreted by the judiciary, judges, and attorneys. Telecommunication lawyers exist to interpret policies, hence individuals and organizations using media equipment and services can interpret media policies to fit their needs. In all systems, however, policies are not static, they must change to meet the needs of their users. In autocratic systems where posts and telecommunications executives make policies without using the opinions of the masses at which they are directed, their implementation is unsuccessful (See Figure 6.0).

In Japan, regulators and providers are one and the same, because

Figure 6.0: ***Information policymaking in autocratic regimes***

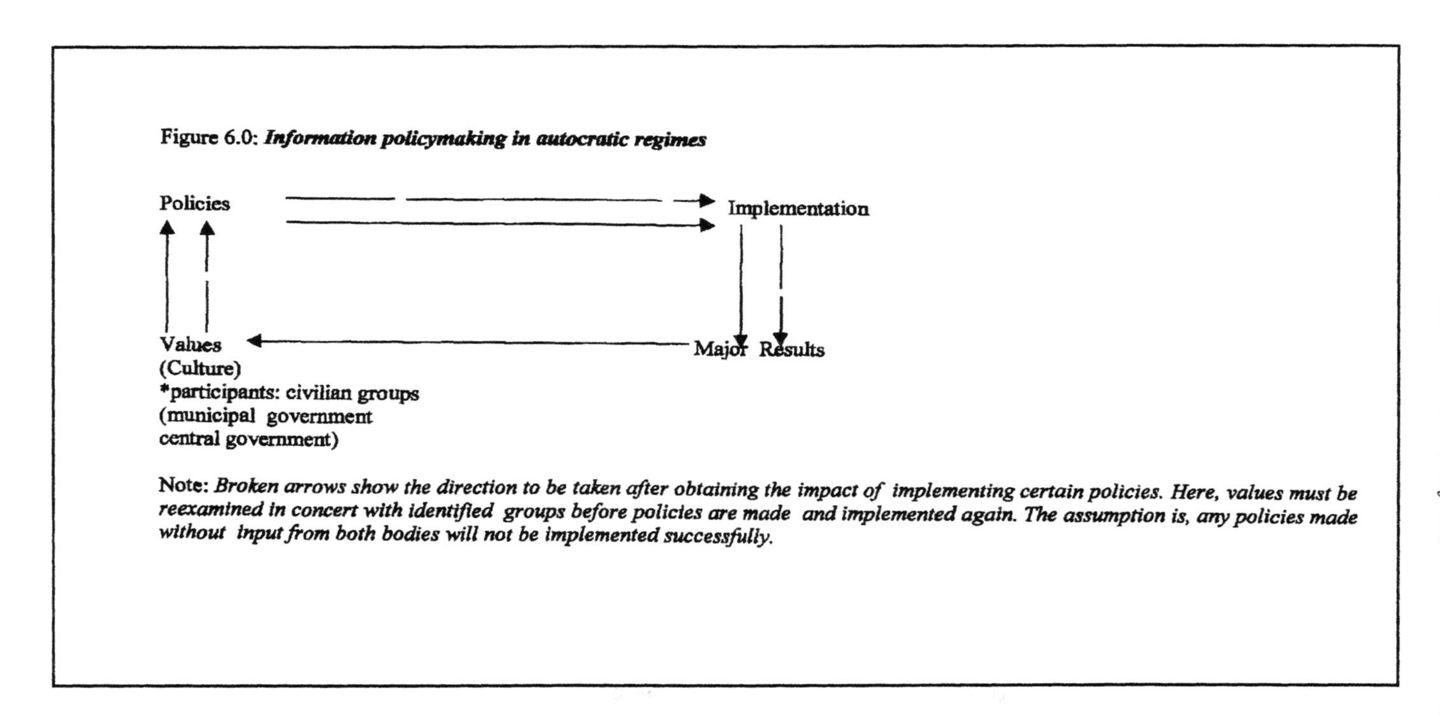

Note: *Broken arrows show the direction to be taken after obtaining the impact of implementing certain policies. Here, values must be reexamined in concert with identified groups before policies are made and implemented again. The assumption is, any policies made without input from both bodies will not be implemented successfully.*

Japan manufactures telecommunication equipment. Japan has been competing with the United States for the provision of telecommunication systems in Africa since the 1980s.

In African public offices, policy-making is inconsistent with the needs of the majority because policies are designed to satisfy the immediate interests of those holding high offices. In *Communication efficiency and rural development in Africa* (1995), the author states that policymakers in Francophone Africa often connive with their cronies or act autonomously. Policies are weak, limited in scope and often not applied because the executives (ministers) are neither trained in telecommunication management, nor do they normally consult with experts before making choices. Also because of unstable political leadership, especially when such ministers are replaced without having mastered or implementing the policies they designed, it becomes difficult to evaluate the quality of policies. The policy statements which are usually speeches and ministerial decrees, are fluid; there is no comprehensive document left behind for research and future policy adoption or analyses. This prevents international telecommunication industries and business entities from carrying out tangible and sustainable transactions with indigenous telecommunication institutions and local markets. Given the status quo, an international organization can easily impose its policies on the latter.

Another reason for the prevalence of bad policies is the nonchalance of policymakers. As mentioned earlier, governments place more emphasis on commercial, agricultural, and political issues rather than on the information industry, despite the latter's potential in facilitating life and expediting development. Most telephones and other postal and telecommunication equipment in Africa were imported by colonial administrators, tourists, and missionaries to serve emerging European communities and other immigrants who had to maintain contact among themselves in the vast land, and with governments, families and churches in their home countries. There were no specific policies, so they could have used telephones and radio sets as they desired. Similarly, local authorities did not normally manage even telephone bills, although they often paid bills for the repair of telephones and radio sets. The non-payment of phone bills to African authorities helped ruin local economies.

Those practices notwithstanding, law was central to colonialism in Africa. Besides serving as essential elements in European attempts to establish and maintain political domination, laws, courts, police and prison camps were instrumental in restructuring local economies to promote the production of exports for European markets (Mann and Roberts, 1991). Legal rules and procedures set forth by the colonials became instruments of African resistance, adaptation and renewal. Basically, African traditional rulers, their subjects, and those working for Europeans underwent three stages of change concomitantly. They resisted European governing techniques, then they were forced to adapt to Eurocentric ideals which eventually changed their world views, and caused them to see such values as superior. When a common African male trained by the Europeans arrested, tortured, and imprisoned a King whom the natives revered, the natives could no longer resist the colonials. Certainly, there was a mystery behind European governance. This mystical image and the vulnerability of local leaders may have generated a context for creating and undermining state policies in subsequent–post-colonial—African countries.

Regulatory Framework in Other Developing Nations

Among developing regions, Latin American countries have been leading in the passing of comprehensive legislation regulating the transfer and control of technology (Radway, 1983, p. 51). Through these regulations, the countries have enhanced their economic objectives, some of which include the maximal utilization of locally produced raw materials to create more employment and reduce unnecessary imports; improve the bargaining ability of technology consumers in these countries and control intra firm operations of transnational enterprises (Cabanellas, 1979, pp. 29-30). Most of these objectives have not been fulfilled following major developments in international business, especially the NAFTA—North American Free Trade Agreement—a treaty signed by US, Canadian and Mexican authorities in 1992. NAFTA oversees free trade practices between those countries. Because the U.S. influenced the development of this treaty, critics of the agreement assert that Mexico has been the major loser because following passage of NAFTA, it had its highest unemployment rate with 105, 225 fewer jobs than in the previous year, and 80 percent of them lost in manufacturing. Pizarro (1994) has pointed that although Mexican exports grew by 25 percent in the first quarter of 1994, and imports rose by 73 percent, the number of manufacturing jobs in the U.S. grew faster than in member countries. The balance of cheap consumer goods, capital goods, sales, and job loss and job creation, will have a negative impact on the economy of Mexico, even though economist McGraw (1996) has predicted a 100 percent rise in trade activities between the parties in the next five years.

The decline in employment certainly defeats the regulatory objective of Latin American countries which is to create more jobs and reduce unnecessary imports. Although these countries have a policy framework that can be adapted by African countries, transferring technology through a free trade agreement with industrialized countries would further slow their ability to advance economically, especially when their technology and managerial techniques are drastically limited. For developing countries to benefit from free trade with developed countries, their governments should privatize industries. Doing so would accelerate competitiveness; more quality goods would be produced for a larger market and the local industries would remain in business.

African governments perceive foreign capital investment and technology as the bases for building their economies. In most of their development plans, which span across five years, they encourage foreign investment and provide plans to safeguard the investments. Such investment would result in a gain in technical knowledge to be used eventually toward the economic development of the country, if properly organized. Generally, cooperation is required to ensure development. Here, *cooperation* means funding participation, corporate partnership, and active participation between country requesting project (home government), agency providing funds, and experts or personnel carrying out project. No project can be successfully planned, implemented, and sustained without consent of home government and a clear description of potential economic benefits to investor, sponsor, and citizens.

Any realistic project would require three strategic preliminary steps by the home government:

1. A needs assessment based on economic and social benefits;
2. Project design (clear proposal); and
3. Contacts with potential sponsors to initiate project study and underwriting.

Potential national benefits should be clearly stated. They should include all aspects of capacity building, especially human investment and financial capital potential. Proposals that describe social support for a free market economy like a national information network—discussed later in this chapter—are likely to contain open-door policy issues. The governments of Kenya, Ghana, and South Africa which had an open-door policy to foreign private capital and technology investment since independence have, to some extent, achieved the following objectives:

1. Created employment opportunities for their citizens;
2. Improved and maintained working/diplomatic relations with the foreign countries of the private companies;
3. Intensified competition in the marketing of products locally, hence;
4. Improved the quality of services provided locally;
5. Improved personnel work ethic in telecommunication industries; and
6. Contributed to the increase of national income by widening the base of the national economy and by providing a wider variety of products, which can increase consumer choice and satisfaction.

Through these practices, country would reach an international status, and it would be able to influence global political decisions. Additionally, the main objectives behind importing technology into these countries have been to curb problems of unemployment, create self-sufficiency in the use of basic commodities, and enable locals and foreign providers earn money through import-export transactions. Despite various amendments in their communications, investments, and customs policies, these and other governments have become increasingly interested in strengthening their economies. As such they have finally resorted to privatizing most state-managed institutions.

Privatizing the Industry

Privatization has become a household term used by African governments and foreign providers to designate reduction of government and maximization of private sector. This change in government policy is taking place about three decades after their independence from foreign rule, certainly the right time for young adult nations to encourage the establishment of a market economy in order to compete with other nations on an economic scale. As no country can influence international politics without having a strong economy, African nations certainly need to trickle down their economy to allow many new industries to become more creative, aggressive and expansive in the manufacturing and marketing of their products.

The incentive taken by governments to encourage private investment in the 1990s indicates an awareness of the New World order that is itself charged with new markets and fast expanding economies. But it is hard to coprehend the way large economies (advanced nations) enhanced by modern technologies are prepared to do macro business with, or enlarge, smaller economies (developing nations). Practically, developing nations cannot have a fair share of global profit by opening their markets to these large foreign corporations because the latter have a longer history of privatization and the former that of indebtedness to the latter.

The idea of privatizing telecommunications in Africa was initiated by the International Telecommunication Union (ITU) at a conference in Europe in 1967. ITU sought to improve telecommunication by suggesting that it could be used in enhancing development efforts; support education, health, and commercial activities in rural areas of 30 sub-Saharan countries. In 1968, ITU received a contract to conduct a study for a pan African telecommunication network—indeed, a business route through West Africa. This contract signed with Acres Ltd. International, a Canadian engineering firm, involved the inclusion of 156 circuits to the existing 230 in order to serve cities and major centers in the region. Twenty-six nations received solar power, which removed dependency on the delivery of fuel. The Canadian International Development Agency (CIDA), has played an important role in advancing steps toward privatizing the industry by providing to national agencies support in developing billing and cost recovery systems, developing an international traffic structure and traffic management techniques, repairing parts, and providing an approach to the purchasing of equipment parts. The activities of these agencies have set the pace for discussing privatization and regulation of the telecommunication sector among African governments.

Are their policies tied to democratic ideals or to socialist principles, given that African leaders have also experimented with Marxist-socialist values?

State policies should be based on the principles of moral sanity and reciprocity for the state to achieve its international objectives and improve its relations with other nations. This need for global interaction stems from the concept that Africa cannot exist in a vacuum or perish in its economic misery. Rather, its policymakers must aggressively seek international support in establishing and sustaining their industries. And that can only be achieved when they network with foreign service equipment providers in establishing appropriate regulations for their industries and users. (See Figure 6.1)

Providers must understand the needs of users in order to educate regulators on how to design and implement policies. Both domestic law makers and foreign providers must undertake the monitoring of policies. Without placing appropriate labels on the equipment they export to users, they should be liable by international courts. Regulators should also be aware of users' knowledge of the product and must have an understanding of the services providers can offer prior to designing and imple-

Figure 6.1: Telecommunication triad for international & local markets

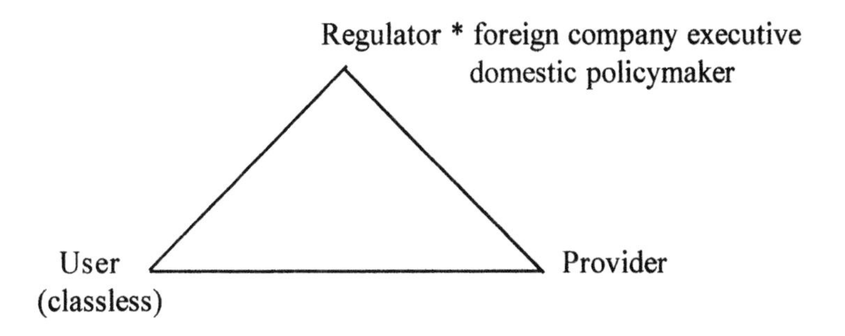

menting policies. Without a triangular understanding between regulator, provider and user, there could be chaos; African markets could fall, Africans might not benefit from the information technology and regulators might be charged for violating the non-interference treaty established by the United Nations, for imposing regulations on a sovereign nation. An effective regulatory body, which seriously considers the needs of the local market must interact closely with the international provider (see Figure 6.2). The user is the most important party in the market because this entity generates finances for advancement of business and increases a country's revenue, not the policymaker. So the foreign provider and local official must exert much effort to serve the user.

That process also entails using state resources: ethnic group values, beliefs, communication experts, and existing communication infrastructure to construct policies that project mutual respect, reinforce social and ethnic values, justice, corporate survival and the economy in general. Any state, which does not design policies based on those factors, cannot effectively implement them because they form the basis for human action. You are how you think, so what you say and do are based on what you think. Putting others before yourself, considering others as being equal to you materially and psychologically should include the

Figure 6.2: Directives for Serving the Local market

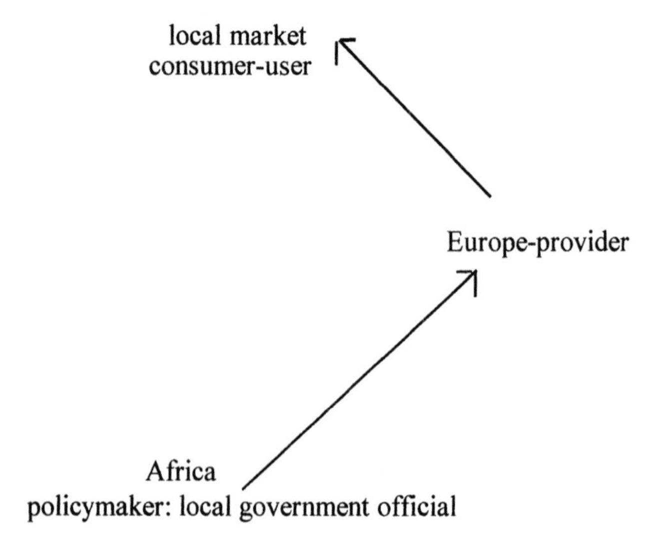

framework for designing national policies. When national policies are designed without including national values, a nation becomes vulnerable to the influence of international policies. Thus, no matter what their intentions, policies should satisfy or meet the legitimate needs of the majority population.

Opinions on this issue differ depending on governing principles of a nation; Africans should not allow their telecommunication institutions to be totally controlled by their governments because they cannot manage the high cost of purchasing and maintaining telecommunication equipment and services with their meager budgets. Moreover, the governments do not consider telecommunication a viable instrument of development, since annual telecommunication budgets are usually among the lowest. Privatization and liberalization of the industry are the logical steps to take, especially with the advent of the New World communication order. The private enterprise will provide creative incentives and generate funds to expand the industry and its services. The governments of Guinea, Kenya, South Africa, Lesotho, and others are privatizing the telecommunication sector. Privatization and liberalization are also important for African nations because they will:

1. Create additional employment opportunities for Africans;
2. Encourage the training the jobless cadres, mainly university graduates, to work in the private sector, which does not practice nepotism, tribalism, and favoritism, practices frequently associated with the government sector;
3. Promote education, because more people will have practical knowledge of modern society and such technologies as computers, facsimile, and TelePrompTers;
4. Advance the use of interactive media which will teach and train individuals according to their ability to learn without the linguistic and technical handicaps encountered in operating videos, instructional manuals and classroom training. (People remember over 75 percent percent of what they see, hear and do, about 40 percent of what they see and hear, and 25 percent of what they see. Moreover, private telecommunication companies could design and implement media strategies for business communications and marketing which could produce significant and incremental sales of products.)

5. Give Africans an opportunity to foster interaction with foreigners, learn foreign culture and acquire state-of-the-art job skills through contact with foreign expatriates and sophisticated equipment.
6. Privatization will also elevate individual competency and competitive skills because individuals will be hired based on competence and experience, not through "back door" politics. Competition for qualified workers will improve the hitherto *laissez faire* work ethic and corruption typical in African public offices. It will increase the quality of work and eventually make Africa more marketable in the economic world.

The industry could improve the quality of life for local communities in many ways. A telecommunication industry with good policies will improve the economy of the local community through asset management. It will help small-scale business persons to improve the way they managetheir assets. Record-keeping is one of the largest obstacles to effective business management in Africa. Records of such important documents as birth certificates, medical history, and academic qualifications are usually stored in paper form and quite often they are destroyed by fire, vandalized, or altered by fraudulent and corrupt government officials. A computer company can offer its services to the government to manage public records.

Also, a strong telecommunication infrastructure should be able to:

1. Enhance economic development opportunities and attract industries;
2. Enhance the development of information infrastructure, e.g. improve services, have machinery do what people do, provide quality information;
3. Bring new companies into the community. Industries need telecommunication to ease communication transactions;
4. Create a regulatory environment and competition among local companies;
5. Serve as an investment code—people should be able to communicate through video systems;
6. Promote equal access to local information infrastructure;

7. Facilitate public access to local information through cable television channels and computers;
8. Encourage broad access to communication by supporting development systems; and
9. Assist local telecommunication industries in the establishment of partnership with other communities and encourage greater dialogue between officials and civilians for community development.

African media practitioners are highly concerned about technology and program content because foreign industries can and do import programs and equipment that can destroy societal values in developing countries. But how can local communities develop policies that meet their needs and propel them into cyberspace? Unless foreign telecommunication companies establish themselves in African regions, who will fund the type of industry that can make the local community to reap the afore stated objectives? The funding of any telecommunication industry by a foreign entity includes traps. Funding is often based on the types of products or agendas the entity intends to promote. Some donor countries worry about implementing policies that infringe on their constitution. For instance, the U.S. Constitution gives people the right to full expression and still denies them the same right by introducing the V-chip (an electronic device which enables certain television programs to be blocked out). Congress has debated the extent to which a local or long distance telephone company may interfere in each other's markets or manufacture new products. Selective viewing and the manufacturing of new products have become serious policy issues in America; meanwhile developing nations are still preoccupied with privatizing the industry and are less concerned about program content.

Telecommunication Status and Policy Update: Country Profiles

Botswana

Located north of South Africa, Botswana is a maleable place for the establishment of a strong communication economy, because:

1. A land and water area of 585, 370 Sq km and a high elevation point of 1,489Km;
2. A semi-arid climate: warm springs, hot summers, no winters;
3. Males and females aged 15 years and over can read and write. The literacy rate is 70 percent in a national population of 1.5 million;
4. Multi-party, multi-racial democracy. Its modern democracy is 31 years old while its traditional democracy is one of the oldest in the world. Also, it hasa legislative branch makes laws. Public officials are elected to office;
5. The country has a $500 million investment and an annual production of 2 million carats of diamonds—the largest producer of diamonds in Africa;
6. A purchasing power parity of $4.6 billion (1996 est.);
7. English (used by most electronic information consumers worldwide) is the main language of business and instruction.
8. A high birth rate.

Given those socioeconomic factors, corporations can gainfully invest in an information network in Botswana. Needles to point out that the stable political climate, strong potential for growth in the economy, high foreign exchange reserves, and a high level of education are the right tools for investment in any developing country.

Because the legislative branch makes laws governing national operations and serves a five-year term, new policies affecting the socioeconomic development of the nation can be easily passed and implemented. For example, the Botswana Telecommunications Corporation (BTC) which was privatized in 1980 has been developing and managing the nation's internal and international services. For almost two decades, the industry has made more progress than those of its regional neighbors—the Zambia and South Africa telecommunication sectors. Based on 1998 estimates there were 50,000 telephone lines consisting open-wire microwave radio relay links and radiotelephone communication stations. The international telephone system comprised microwave radio relay links to Zimbabwe, Zambia, and South Africa, and one INTELSAT operating out of the Indian Ocean. There were 13 FM and 7 AM radio stations. Since 1996, about 40 villages have been provided telecommunication services and the government has invested about 250 million pula on a microwave ring to cover 4000kilometers nationwide. In Gabarone, the capital, Orapa and other urban areas with high population density, fiber optic cable was used to diversify, reinforce, and facilitate the provision of data services and telephone conversations. Even the computer industry has become competitive as over sixty major dealers from around the world provide PCs and related products in the Botswana cities. The country produces a monthly computer magazine that updates, for local users, developments in the computer industry. Also, an electronic billboard allows access to about 1,700 computer-related topics. With such improvements, foreign computer dealers companies can harvest an estimated profit of 30 million pula per year while providing more jobs for Botswana residents.

The operating information technology and other new infrastructures in Botswana have made the country one of the fastest growing receivers of foreign products among such highly populated and technologically advancing nations like Nigeria and South Africa.

Chad

The growth of Chad's economy has been deterred by natural and artificial forces; drought, food shortages, civil war, harsh governance, and conflict with other African countries. Its GDP per capita was $600 (1995 estimate). Since 1986, the GDP level has been based on cotton which accounts for 48 percent of exports. Between 1986 and 1990, over 80 percent of the workforce was employed in fishing, and since 1991 subsistence farming has been the people's primary work order. The country's national product per capita was $215 million with revenues yielding only $115 million while annual defense expenditures ran about $60 million.

With a land area of 1,285,200 Sq. Km of which 13 percent is arable, and an estimated population of 14 million (1997 est.), Chad only has 15 kwh per capita of electricity supply and a 40,000Kw general capacity. This means that agriculture, cotton, textile, and other industries do not have enough electricity to effectively conduct business and that an electronic communication network cannot fully operate. In addition to low power supply, transportation systems are limited, severely hindering national development. The highways only total about 35,000Km with 7,500 km of gravel and laterite and about 40km bituminous. Of the 69 documented airports, only 5 have permanent surface runways and one capable of receiving large aircrafts from Europe and other regions.

The country is run by a military regime with about 4 million workfit residents aged 15-64. About 1.5 million males aged 15-49 are fit for the army. Appointments to government offices are made through a presidential decree, meaning national policies can change arbitrarily, and the cabinet can hardly be receptive to long term plans. The quality of education and patriotism and the rate of sharing progressive ideas among groups and institutions have declined since Chad's independence, partially because of unstable governance. Consequently, the capacity building level is low.

Since its inception over three decades ago, the telecommunication sector has been virtually ineffective. With about 8,000 telephone lines (only 1 telephone line per 125 people), the extent of sharing or managing information within and beyond national borders is limited. In addition, natural forces have hindered human capacity. With a hazardous tropical climate, hot and dry weather and dusty harmattan winds, drought, sporadic ethnic rivalries, plagues, and famine, the idea of installing and managing a national or regional telecommunication network for efficient business transaction does not appear feasible. Despite the poor climate and past political and electronic communication problems, the government has shown signs of improving telecommunications. The Universal Postal Union and ONPT have been working together to increase the number of long distance lines and reduce the costs of telephone calls. In 1998, officials held discussions with World Bank staff to restructure the industry. The discussion included a plan to liberalize it and develop an information network that would bring more competition. Moreover, a system of radio stations has been set up for inter city use. Stations broadcast on one FM and 6 AM stations and listeners receive international programs broadcast daily on two channels: Africa No. 1, a regional network originating from Gabon (which covers French-speaking listeners in Central and West Africa), and *Radio France Internationale.* Viewers in cities bordering Chad, Cameroon, Nigeria, and Niger receive international television broadcasts periodically, partially because they have access to INTELSAT, the Atlantic earth station. The governments select a limited number of international programs (programs from Europe), and encourage local origination. International programming includes sports and politics.

Despite its unstable economy, Chad has the capacity to become a major information marketplace in the 21st century, in terms of human capital. The fertility rate is 6 children born per woman, life expectancy rate 53 years, and a population growth rate of 2.9 percent. Using this data, telephone companies, for instance, can eventually maximize consumership by providing different products and services to different population segments at an affordable cost.

Guinea

The Guinean government allocated 470 million to privatize the Post and Telecommunication ministry. The *Société des Télécommunications de Guinea* intends to:

1. Rehabilitate, modernize and extend the telephone network in two main cities;
2. Implement 14 automatic telephone systems;
3. Complete telephone installation in most parts of the country by the year 2000;
4. Put into service all phone lines with neighboring countries; and install a coast station in Conakry, the nation's capital.

These activities will promote economic communication in that Muslim country which has traditionally regarded disseminating information for economic expansion as a thing for the privileged class alone. Such activities would revive the waning communication between Senegal and the Middle East, as more Muslims would be able to use voice systems to interact with each other and hence, enhance their religious and cultural ties. While the world's audience has depended on the Western media for a balanced coverage of world news and events ("balanced" refers to stories on disaster and success), it has covered the Islamic world with apparent malicious intentions. From the 1980s, beginning with the Salmon Rushdie saga and the Ayatollah, to the Gulf War and Saddam Hussein in the 1990s, viewers have been negatively influenced about Muslim leadership and Islam. The Muslim community has not been able to defend its image internationally through world media because it lacks a global information technology. That Western electronic media has depicted events in the Middle East, Libya and other Arab nations in a negative light

(Mowlana and Kamalipour, 1994, Mowlana, 1996, Jin, 1991, Luostarinen 1991, and Kazan, 1993) and destabilized the region, justify the need for the establishment of an exclusive electronic information network in the Middle East. With the installation of a telecommunication infrastructure in that region and other Muslim nations as Senegal, the Muslim community would eventually disseminate Muslim propaganda and regain its place on the world's political stage.

Kenya

According to a 1991 report published by the Central Bank of Kenya, the telecommunication system in Kenya was deliberately structured to serve the European settlers and colonial administration. However, Kenya has become the most powerful post and telecommunication country in East Africa, because of its central geographical location. It has served as the headquarters of the East African Railways, road, and telecommunication. The Kenyan government has since taken that advantage to develop, expand, diversify and modernize its services, this had made Kenya an international center for transit rail, road, air traffic and for telecommunication activities throughout East and Central Africa.

Kenya has one of the most advanced telecommunication systems in East Africa. With a steady increase in the call completion rate from 53.6 percent in 1986/87 to about 83 percent in the late 1990s, and with over 4,000 coin telephone units in operation, Kenya could become Africa's front-runner in the information superhighway along with South Africa. According to the 1987 Annual Reports and Accounts of Kenya's Post and Telecommunication Corporation (KPTC), international telex calls rose by 1.5 percent (1,292,331) in 1987-88 fiscal year compared to only 1,272,831 in the 1986/87 period.

KPTC, a government owned parastatal agency operating under the Kenya P and T Act runs a postal, telephone, radio and other telecommunication services nationally and internationally. KPTC whose objective is to expand its network, modernize and diversify its services in the rural and urban areas in order to stimulate socioeconomic development (Kenya Land of Opportunity, 1991, p. 183), is fulfilling its mission. Telex subscribers increased from 140 in 1977 to over 4,000 in 1994 and telephone subscribers increased from 5,681 in 1987 to over 25,000 in 1995. Between 1983 and 1988, the volume of inland telegrams rose from 1,000 to 1,447 and that of paging messages also rose from 50.8 percent to 78.7

percent. With 20 telecommunication companies providing block wiring and two international systems located in such high revenue areas as Nairobi and Mombassa the chief commercial city, and with an exchange capacity of over 400,000 telephone lines and a projection of 1.5 million telephone lines by the year 2000, Kenya's digital automatic telephone exchange system will rank among the best in Africa. However, a capacity of 2,000 subscribers is rather limited compared to the country's population and UN standard of one phone unit within 5 km in all regions of Africa. This slow increase is commensurate with the pace of the economy. Basically, telecommunication traffic may only increase depending on an increase in services or an increase in the demand for electronically processed information. Since the volume of transmitted paging messages increased from 81,989 in 1987 to over 100,000 in 1995, it is clear that people living in Kenya have found electronic communication beneficial.

While there has been no comprehensive document or data which identifies business calls from personal ones, the use of telephones, facsimile, and pagers has been increasing. By 2000AD, there will be one pager per 50 persons in urban areas because its price has been dropping. Additionally, more providers and vendors are becoming available nationally and internationally. Meanwhile, in the U.S., usership of this system is no longer restricted to the privileged class; doctors and lawyers as more common people are now using it to keep in touch with their loved ones. As of 1998, over 10 million pagers were in use in the U.S. were in use and about 40 percent of the market were non-professionals—youth. Between 1998 and 1999, drug dealers and lovers ranked among the majority. The reason for the availability of pagers is its low price. With $US49 down payment and as low as $10 per month rent charge, a user can maintain paging services for years. On the other hand, common people in Kenya and other African countries cannot afford to spend that amount on information services for several reasons: limited employment opportunities, and a general laxity in investing on information. Because the majority of people obtain information from and rely on such such secondary sources, the desire to invest on electronic information sources is limited to business people, high level professionals and high profile personalities. That is partly why there is slow progress in the making of telecommunication policies in Kenya.

While the phone sector is making minor inroads at improving the telecommunication industry and services in Kenya, the Kenya broadcasting Corporation KBC which controls the broadcasting of receiver licenses and telecommunication equipment repairs, has been receiving satellite pictures through the Longonot earth satellite since 1984. Additionally, satellite has 14 television and radio broadcast centers that permit the quick dissemination and exchange of electronic information throughout Kenya.

Lesotho

The telecommunication policies in Lesotho, one of Africa's remaining kingdoms, may become incomprehensible as new political activities befall it. With the monarchy threatened by a democratic process, the hitherto autocratically-designed policies are being tested. Based on a treatise published by the Department of Customs and Excise, 1978, the minister has the powers to amend the importation of telecommunication equipment as deemed expedient in the public interest. According to Vol. VII, Rule 2a, Number 10 of the 1982 Customs and Excise Act, the local Assembly ruled that any tariff applied to an article cannot be changed when the latter is imported in an incomplete condition. This implies a foreign telecommunication company could import an unfinished or poor quality equipment and still pay the same tariff charged for finished or good quality equipment.

However, under Section C2, Number 2, protection is accorded to a new industry for eight years or less, except with a prior consent of contracting parties. Based on that clause, a foreign industry in Lesotho can be protected for eight years and during this time it can enhance its service to the public and hence stay longer. But because such articles as 11, Number 3 concerning export/import matters have allowed the monarchy to restrict the importation or exportation of goods and services, the risks of improving the telecommunication industry may be high, despite Lesotho's reputation as one Africa's richest and least technologically advanced nations. While economists working for major world organizations may consider these policy issues an attempt to provide a yardstick for evaluating the quality of import goods, they do not support the advancement of local industry incentive. Any nation which restricts the flow of foreign products without the capability of producing quality products locally, reduces its chances of competing on the international level.

Although the Lesotho government has privatized the telecommunication industry, it has not placed much importance on the company's development. Instead, the government has invested more on external trade than on telephone systems. Between 1983 and 1986, there were only 5.7 telephone lines per 1,000 people, increasing in 1987 to 17.9—28,583 telephones—compared to 845.9 million maloti ($1US = 2.5 loti) on imports in 1987 alone. Even with a steady increase in the use of telephones from 9.1 per thousand in 1985 to over 20.6 per thousand in 1995 following the government's decision to open its doors to foreign markets, (especially the U.S.,) one should remain skeptical for the following reason: the monarchy may seek to consolidate authority by restricting the importation of telecommunication products and services. Nevertheless, through new diplomatic relations between the U.S. and Lesotho and Lesotho's invitation of American businesses in recent years, new policy ideas may be generated that would increase the importation of advanced telecommunication equipment and personnel.

South Africa

With a new government in power, there is an intense focus on improving local and international communication. The new government has been developing telecommunication policy issues which when carefully screened and implemented could place the nation abreast of industrialized nations in terms of supporting electronic information circulation and business transaction. The rapid development of the telecommunication market calls for clear policies, however, one of the consequences of the country's political transition is that the telecommunication sector has been subjected to intense scrutiny. Since policies affect future options, South African authorities have undertaken intensive research to determine the implications of, and options for, its participation in the evolving global information society. In its Green Paper published by the Ministry of Posts, Telecommunication, and Broadcasting in 1995 which is also available in Afrikaans, Zulu, and Sotho languages on the Internet and in other electronic forms for public comment and suggestions, writers purport:

> A framework concerning information in South Africa should cover a wide variety of issues, such as freedom of access to information, privacy, intellectual property rights, cultural development, the development of viable

local information, and information technology industries and the development of sectional information systems to meet social objectives.

That policy framework can contribute to a future information policy that would advance market structures, demonopolize the sector and increase the country's chances of enjoying the privileges of a free market economy and free access tothe global information market. When new strong policies are implemented, those who have never had access to minimum information services, especially in the rural areas, would have a chance to obtain information about their country, their government and the world. However, industry restructuring towards universal services raises several policy issues: How would the objectives of universal access be defined and by whom? How would quality service be evaluated and/or offered? What would be the rate of service provision? Could universal access be provided at special rates? If so, at whose expense? Would there be long-or short-term provisions for special markets, e.g., rural residents and, for what reasons?

Since the sector is being structured around Telkom, the basic provider can realize advantages that include the establishment of standards and other economies of scale. However, conditions for supporting monopoly provisions have changed because new technologies have lowered the previously high entry barriers in many services and thus, have undermined the potentials of a monopoly to maintain itself. Cross-subsidy practices are no longer acceptable, because international competition requires tariffs be based on actual market rates. Thus, to ensure that fundamental objectives of a policy making entity are met, that entity should:

1. Have an explicit service provision for South Africa (SA) which should include conditions for obtaining service or vendor licenses;
2. Indicate when, where, and how finances should be paid to maintain the operation of the institution. For example, there should be a provision for local vendors to pay a smaller operator license than foreign vendors. This would enhance the local economy—more South Africans would be employed—and would limit foreign ownership and investment in SA, through the latter; and

3. Find qualified personnel to manage the funds and make decisions on global service matters.

Telecommunication ministry officials should be grappling with the above issues, including whether foreign investment should be sought and how investors' confidence can be insured. In a country which has just been emancipated from decades of apartheid, finding the right policies for a sensitive sector as telecommunication can be difficult. South African authorities need to design and implement comprehensible policies to monitor the influx of foreign providers and services and prevent an economic apartheid. As a new independent nation pursuing democracy, the government should work with local, national and international experts in developing policies that would lead the country along a high economic road paved with electronically processed information into the new millennium. The government might become vulnerable to offers from rich nations by not closely examining blue prints like high interest rates and sociopolitical influences to be exerted by the foreign rich power. It is rather early for South Africa to import all forms of information technology or implement policies that encourage such activity because SA is on the crossroads politically. Information systems made available to the public may also be used by opposition parties to step up their aggressive behavior against the ruling government. Opposition parties and social deviants might access and use information to upset the social norm and destabilize the government. Article 5.8.1-5.8.3 of the Public Sector in the White Paper deals with the steps needed to restructure information and guarantee the peoples' right to know.

Despite such conditions, South Africa has the most advanced telecommunication infrastructure in the continent. Out of a national population of 30 million, over 5.3 million have installed phones, which accounts for 39 percent of the total lines installed in Africa. By 1995, there were about 90 lines per 1,000 inhabitants, as opposed to 1 per 1,000 in other countries. South Africa's government-owned Telkom system has processed over 26 million call units each year and about one quarter of the completed calls have come from abroad. According to the 1994 South Africa Yearbook, an average of 3.45 million calls completed per month were conveyed via 2,588 satellites and 1,228 submarine cables. At this pace, half of the nation's population should have telephone lines by 2000AD. South Africans could use a toll-free number to dial back home from over 15 different countries, through the South Africa Direct program. Even residential customers have access to a confidential toll-free service and which has automatic facilities. The call completion rate has been facilitated through MAXINET, wherein the caller does not use dialing codes. Although South Africans use 10-digit numbers like American residents, participants calling in during a national contest have a higher success rate than those using dialing codes. Business people with one toll-free number may have calls rerouted simultaneously to more destinations in South Africa than elsewhere. Additionally, South African motorists have the capability of making phone calls worldwide from their vehicle.

Despite its advanced stage, the South Africa telephone market is smaller than that of the US. While the four major U.S. telephone companies have numerous customer service centers to handle subscriber accounts, billing, installation and other queries, South Africa had only one station in Pretoria in 1995. The nation should have 166 centers by the beginning of the 21st century.

The Ministry of Post and Telecommunication should restrict access to secret government files and white papers, until such material no longer poses a threat to its security. By making its policy proposal available to the public for discussion (the July 1995 Telecommunication Green Paper), the Ministry may have lent itself to heavy public criticism. The suggestion that international input on ownership and investment can be sought may have upset a majority of South African citizens who fear undergoing another apartheid ordeal. The question on Page 12 of the Green Paper : "Should there be any limits set on foreign ownership" (of telecommunication sector) suggests they are capable of relinquishing authority to a foreign body.

Assuming a foreign body controls the telecommunication sector, South Africans would have limited control over which type of equipment to be imported, its tariff determination and legal issues regarding its quality. The open door mentality expressed in the Green Paper is therefore inappropriate for the determination of telecommunication policies because they are ambiguous like the First Amendment of the American Constitution, which guarantees freedom of speech but also restricts certain press activities and the use of obscene language. But if the South African telecommunication ministry must emulate the pattern of formulating policies from such advanced democracies as the United States, it must realize that American telecommunication policies are partially based on the multicultural background of its residents and that court systems and in formation consumers constantly grapple with interpretation problems. For example, journalists are liable for withholding certain information from the public--or for libel; pub;ishing information that can damage a person's image, hold him up to public scorn, ridicule or embarrassment. The South African telecommunication policy makers must carefully select the language of telecommunication policies.

Zambia

Telecommunication regulation in Zambia has been changing since its government initiated its privatization in 1975, by making the General Post Office a parastatal organization. Thereafter, the Post and Telecommunication Corporation (PTC) became a part of the Zambia Industrial and Mining Corporation (ZIMCO). Under this new organization, govenrment's investment would be limited and the company would operate without subsidiaries (Akwule, 1990). However, communication regulation became a significant issue in Zambia only after the Privatization Act of 1992 and the creation of the Zambia Privatization Agency (ZPA). The PTC has since become a parastatal enterprise operating under new major regulations. Long term objectives of privatizing the PTC (known after the 1994 Telecommunication Act as Zambia Telecommunications—ZAMTEL) include; the issuance of licenses to cellular operators, the provision of access to funds for financial markets seeking competition, and the designing of appropriate staffing policies. According to the 1994 Telecommunication Act, telecommunications would be regulated by the Office of Communications (OFCOM) modeled after the UK's Office of telecommunications (OFTEL), and the U.S. Federal Communications Commission (FCC).

The emulation of foreign regulatory techniques may be dangerous considering political climates in foreign countries from which such policies were made introduced differ from those in Zambia. Zambian authorities must realize that Zambian telecommunication consumers are not as sophisticated as UK and American counterparts in their observance of telecommunication policies because policies in those countries were generated by specific historical circumstances. Hence, anyone attempting to model from such systems must also take into account its peoples' legal history. A country which only became independent recently, Zambia might not be able to effectively adopt most of the policies modeled from Anglo-Saxon and American judicial systems. Moreover, the multi-ethnic differences and political malaise that often threaten the nation's stability may continue to rise if Zambian telecommunication authorities do not consider ethnic sensitivities when making policies. Apart from nepotistic tendencies, the fear of corruptive practices that hinder the fair implementation of the policies cannot be ignored. Some officials in the Ministry of Communications and Transport have had a vested interest in seeing a particular foreign company win contracts to provide cellular phones and services in Zambia.

A review of its recent regulatory system shows the Ministry of communication and Transport, formerly PTC, has been focusing on providing short term licenses for providers. A temporary license was issued to Jula Communications, a local firm, to import and install telephones where phone cards could also be used. Although the firm pays ZAMTEL 10-15 percent in royalties, it has to share its profit with the foreign provider, which may stall the local economy. Given the suspicious atmosphere surrounding the credibility of telecommunication regulators, bids and proposals for licensing may not be processed under fair conditions.

Similar problems facing the industry have been generated by a decision from Zambian authorities to privatize it. The National Union of Communication Workers (NUCW), a part of ZAMTEL , which advocates for employees' concerns has been concerned that most people would lose their jobs and nepotism would increase following privatization. They have also been concerned that their position in the industry would be weakened and their opinions on important issues ignored. The fear of uncontrolled trading practices is also legitimate; providers could influence local policymakers and customs officials by negotiating private deals for the importation of their products. Manpower, the most available resource which contributes to the growth of the local economy would be replaced by machinery. In the PTC, there are poor revenue-collection practices, insufficient tariffs levied on consumers, fraud, unmotivated underpaid workers, and under-investment in the entire industry. In 1995, the government was reluctant to establish an independent regulatory body, probably for fear of relinquishing authority to that body. The government was not able to manifest its authority under sections 5,6,7, and the 1994 Telecommunication Act which dealt with the award of licenses to operators.

The Zambian government has not clarified its role regarding the privatization of the telecommunication industry. It is doubtful what role it will play if privatization actually takes place. It has been depending on Japanese bilateral aid for its development. Moreover, because of government's limited commitment to the sector, difficulty in maintaining the systems in most areas, low consumer income, and poor access to subscribers due to poor road conditions, Zambia like other African countries shall continue to face problems in developing the industry, or providing telecommunication services. These problems partially account for the governments' slow pace in privatizing the industry. The point here is, since economic conditions like poor roads, low income and system maintenance require the use of large funds and employment of trained personnel, any organization that the government permits to undertake such responsibilities may eventually influence residents in the region. Residents would become subservient to the authority providing them technical assistance.

Further Discussions and Suggestions

General Policy Framework

Since most African governments are unstable, it is imperative that they carefully make policies that would help them in stabilizing their countries. Assistance from the World Bank and other international organizations should only supplement the efforts of African telecommunication officials. The policy of nurturing indigenous growth through financial and technical assistance should only be implemented when requested and determined by a pro-active policy which can lead to fast and effective change. For an African government to receive financial and/or technical aid from a world donor institution, it must adopt communication policies which can support the nation's economic development agenda. Similarly, the donor body should have pro-active policies wherein only the nations with progressive, results-oriented policies are eligible to receive funds. In a Green Paper prepared by World Bank staff, (Baranshamaje et als, 1995), principal elements needed for designing broader comprehensive telecommunication policies for African government were proposed. The staff suggested the Bank:

> Conduct an intense policy dialogue with national governments to emphasize the importance of taking advantage of the information revolution to accelerate economic and social development, as well as the need for regulation, privatization competition in the telecommunication sector. Augment the supply of Internet services through IDF grants; grants from the prospective InfoDev fund recently authorized by a G7 Ministerial Meeting in Brussels and the Bank's Senior Management; and loans, credits, and investments of the World Bank Group, either as private sector development projects or as components within other operations. (p. 15)

These conditions clearly indicate a strong need for the Bank to create and implement policies that can lead to quick progress and independence while avoiding unnecessary bureaucracy. However, the indication that the Bank can provide grants or co-finance information projects is problematic. Establishing a full Internet node with a server and VSAT (Very Small Aperture Terminal) in a given African country may cost $500,000 or less with an annual operational cost of about $150,000. It is certain that most African countries have not been able to afford Internet services. Therefore, the staff 's observation that a bank loan of $30 million could enable 47 African countries to have full Internet service (p. 16) should be considered. Another observation not mentioned in that report is, as a resource-lending agency, the World Bank encourages governments to use their own money to buy investment. As opposed to the UN, CIDA, and other Canadian and European agencies which diminish the amount of investment a developing country can have by providing it with direct funding and service, the Bank advocates a self-help approach. Thus, Bank investments are less costly and more manageable to developing countries. If the World Bank develops strategies for projects by involving financial institutions, home government, and qualified personnel, such governments would be able to sustain internationally-sponsored projects. However, that would not help the countries, it would only make them more indebted to the Bank, especially when the Bank fails to sustain the project as with other projects over the decades. The real challenge, however, is not finances or debt crisis but organization and political maneuvering. Because of regular power outages, low electricity supply, unreliable telephone circuits and poor services, corrupt government officials, and poor work ethic, funds allocated for such projects are often irresponsibly depleted or embezzled.

Additionally, vague regulations restrict access to information and to the expansion of knowledge. African governments posit that the less informed the people are, the greater government's control of their minds. Thus, establishing policies that give the public free access to purchasing and/or using new information technology eventually weakens government influence. These problems can be overcome if donor institutions implement terms for providing loans, one of which should include supervising projects for which loans have been given. The Bank should also have its own supervisor on site to evaluate personnel work ethic, operation of technical resources and record-keeping (revenues and expenditures). Although some governments may regard this policy as an infringement into a country's internal affairs, it ensures productivity and accountability and offers the Bank terms for providing future loans. Only after African governments have demonstrated maturity in managing their own resources should world lending bodies give them unsupervised loans.

To begin managing their own resources in an efficient way, however, government officials should collaborate with media experts in designing and implementing information programs that educate people on how to use communication channels for self-reliant development. Each government should provide funds for programs that promote ideas of a free market economy, responsible capitalism, and multi-cultural enlightenment. Additionally, each government should collaborate with senior experts in organizing a series of training sessions for media experts, training on how to implement campaign strategies for different aspects of community development. For example, a primary education campaign strategy should involve the expert, a government-funding sector, and campaign tools. Here, senior experts should provide government officials a rationale for training communication experts in order to stimulate government interest and justify funding. This approach can, indeed, sustain the funding of such campaigns and facilitate a positive public response. A sample campaign proposal as seen in the following segment highlights this explanation.

Table 6.0

Seminar for Communication Experts on Primary Education Campaign Strategies

A Proposal Addressed to the Ministries of Education and Information.

BACKGROUND: PROBLEM OF EDUCATION

Practical education is the cornerstone of development in all world regions. It is particularly necessary to Third World countries as they struggle to catch up with the rest of the world socially, economically, and technologically. Global movements affecting world economics such as information superhighway justify the need to increase a campaign for primary education in Africa. Moreover, the twenty-first century is an epoch which requires nations to be self-sufficient technically and economically. General knowledge on, and the ability to, apply global ways of thinking will characterize human functioning and the individual's effective functioning will be based on his/her level of technical education.

However, the level of primary education in Africa, especially the West and Central regions has dropped drastically in the last decade for the following reasons:

1. Lack of motivation among school-age people and other target groups; children, young men and women, parents, guardians, and sponsors;
2. Increase in school fees and inflation. Also, devaluation of most currencies;
3. Target groups struggling for survival. Parents have withdrawn their children from school and have engaged them in petty trading, marketing, farming, and prostitution. Some school dropouts have become gangsters. Only 250 out of 1,000 pupils finish primary school, and about 140 successfully graduate from primary school with a certificate;
4. There is a large sense of hopelessness. About 50 percent of primary school graduates do not attend secondary school because hey find primary education as a waste of time and money, since jobs and successful completion of further studies are not guaranteed;
5. Another cause of low enrollment in primary education include, a diminishing moral and financial support from parents and guardians.

While governments in industrialized regions have successfully campaigned for and implemented different educational systems for the youngsters in their countries, African ministers of Education and Culture have failed to carry out effective campaigns to promote primary education (PE) due, in part, to the lack of faith in the role communication plays in increasing enrollment into primary schools, and to the absence of an effective training program for communication experts.

SEMINAR OBJECTIVES

Given the above conditions, a one-week training program for communication experts from the Ministries of Education and Culture is necessary. The purposes of the seminar are to enable the experts to:

1. Adequately inform target groups of the need to improve primary education;
2. Understand different techniques of launching a PE campaign;
3. Practice techniques of using different communication systems in promoting PE;
4. Understand ways of negotiating with and/or persuading ministers to allocate time and funds for an effective public relations/primary education campaign;
5. Understand ways of managing the campaign over time (months or years); and
6. Practice techniques of managing time and of approaching target groups; government officials and prospective pupils.

There can be no realistic increase in pupil enrollment in primary schools without the input of communication experts because they have the ability to reduce ignorance, misinformation, and misdirection, which characterize the daily operation of rural and inner city youngsters and which propagate social malaise. Hence, experts need to learn strategies of successfully persuading youngsters on the partitioning and allocation of their time. Additionally, since most of the youngsters belong to a religion or observe some kind of ritual, experts should be taught techniques of helping the youngsters in negotiating their "social" (personal) time with "formal education" time.

TARGET GROUPS

The groups to be reached by the experts include; young women and men in rural communities, inner cities, highly populated areas, teachers, parents, traditional, local government authorities and local healers, for the latter are trusted sources of information. Over 70 percent of young people—between 6-21 years—in rural communities are poor, docile, fetishistic, and religious. Through their upbringing they are loyal to the elderly, the wealthy, local administrators (village heads, municipal authorities, uniform officers), and traditional doctors. Communication experts can identify such entities and use them as "information channels" for the collection, packaging, and dissemination of messages that can attract youngsters toward receiving primary education.

METHOD

T he seminar lasts six days. There is a morning and an afternoon session punctuated with one break and one lunch per day. The seminar is carried out in a chronological order, meaning there is a progressive method of training and learning beginning with an overview of communication and the target group and continuing with an understanding of the techniques of using communication systems toward acquiring successful results. Instruction administered by a consultant.The consultant intends to apply the triangulation procedure to ensure an effective communication of information to participants. This procedure is included using electronic equipment, involving participants in the production of ideas, and using instruments to test the extent of their learning. A television set, slides, audio/video tapes, questionnaires, test material, and active participation (question-answer sessions, comments, discussions, and testimonies on field experiences by experts) are among training techniques used at the seminar.

The consultant also trains experts on how to use teachers, parents, healers, the clergy and local authorities, as "information outlets" to expedite the PE campaign because such groups have considerable influence over the lives of youngsters. Specific kinds of public gatherings, church worship services, theatre, and music and other interpersonal models introduced to seminar participants. Experts receive training on new ways of using television and radio time, newspaper and bulletin space, participatory videos, and other agents of mass communication for the campaign.

EXPECTED RESULTS

The seminar prepares communication experts to:

1. Overcome uncertainties and difficulties with the PE campaign;
2. Improve working relationships between communication experts, target groups, and government officials;
3. Bolster respect for and confidence in law-makers' understanding of the role communication plays in promoting primary education and in fostering national development.
4. The long term effect is, experts would be contributing to the growth of education by campaigning for an increase in enrollment in primary schools. That effort would eventually reduce the rate of idleness, urban violence, hopelessness, unemployment, illegal activities (like street hawking, and bribery), murder, stealing, and other youth-oriented foibles which plague urban communities and make the work of public officials more difficult;
5. A successful media campaign which increases enrollment helps prepare youngsters for 21st century challenges which include information power and knowledge on global issues.

SEMINAR SITE

Campaign training site should be selected based on geographic location, easy transportation, and easy access to media and/or communication facilities. A typical schedule for this type of campaign is further provided in Appendix III.

Investment Policy Framework

As mentioned, team effort can ensure the establishment of a national information network. This social approach to national investment is critical given that institutions with the capacity to provide large financial loans, e.g., World Bank, believe that national projects are best implemented through funding participation. Business entities, (international and local corporate executives) government officials, and field managers are to work together to ensure a successful implementation of national or regional projects.

A national information network set up in towns can promote free market ideas and activities all over the country. Information centers in rural areas where the greatest effects of poverty are experienced can reduce such malaise. The centers would operate effectively if a number of initial steps are taken. These steps involve joint participation of international institutions, local leaders, and local firms. The World Bank, potential sponsor of the network, should organize seminars to find ideological solutions. Seminars should serve as a beginning point for networking between the government, private firms, influential local leaders, and international sponsors.

Secondly, the government should create an entrepreneurial strategy for the management of the network. Here, the development of rural participation would require strategic cooperation between national macro enterprises and local micro enterprises. Because they are revered by the inhabitants, the support of politicians, tribal chiefs, and religious entities (fetish and non-fetish priests) is necessary for the creation and sustenance of rural information centers. Next, the company managing the network should offer contracts to international organizations that are fit to provide quality service and equipment to rural centers in order to prevent corruption, tribalism, and nepotism among local employees. Additionally, an evaluation of foreign providers by independent local firms would guarantee transparency, objectivity, and sustainability of the project. There should be two kinds of foreign providers: information brokers and equipment providers. Information brokers should have the ability to manage information; develop, process, store, and disseminate information for socioeconomic development in rural areas. NGOs with expertise on information management should be stationed in areas at initial stages of the project to facilitate transition of the rural residents from an agoraphobic information age to an electronic information age. As rural residents, they are in the best condition to articulate their own development needs which include kinds of information to be stored and/or disseminated and the quality/amount of equipment needed.

Companies contracted to provide equipment to rural areas should consider the knowledge-level, degree of interest, and financial condition of local consumers. Lastly, private profit-oriented firms should run information centers in urban areas where corruption is widespread. Because their survival depends on the profit they make, their tenacity, and other business tactics, private firms are more likely to be accountable and fair to their consumers than the NGOs.

Eventually, all parties would benefit from this diversification approach. For local business executives, investing in a National Information Network (NIN) may create higher profit-making opportunities. By serving as underwriter and negotiator for the chief sponsor and home government, the Bank could restore its public image. The home government would have a chance to make its programs and policies accessible to its citizens.

Although some African governments have demonstrated interest in co-sponsoring projects in order to provide jobs to their citizens, their investment in a national information network has been limited. To increase investment, officials in the Ministries of Finance and Communications should:

1. Contact foreign investment-oriented organizations with proposals that specify the need to establish long-term economy-developing projects. The proposal should show how the project would bring about the country's economic independence. It should specify a *modus operandi;* budget, number and qualifications of personnel for the project, site, duration of project, extent of government involvement, and expected benefits for sponsor(s) and home country.
2. Upon receipt of the proposal, qualified Bank staff should immediately evaluate it based on merit. Then, staff should contact corporations doing business in the region requesting the

project, or seeking business ventures there, and file a report to the home government before the latter lose interest.

3. Thereafter, Bank staff should organize individual meetings with government officials, foreign and local corporate executives, and communication consultants to understand their evaluation of the proposal before having a meeting with all parties to plan the operation of the project. The "operation" meeting should include major local politicians and PR practitioners who can convince the local population (potential subscribers and others) to become involved with the project, as consumers.

In Chad, for example, Exxon, ELF, and other major corporations in that country can sponsor the installation of a national information network. With its large computer network, Exxon can serve as a broad base to the community by establishing computer centers in different towns, and provide training, free computers, and computer services to Chadian residents for two years. During the third year, Exxon should charge users an affordable fee. Such investment would provide long-term profit for the company. Also, the government should set up rural information centers and invite corporations to buy its services or levy taxes for the use of its services. Through such efforts, the financial and technical support of the corporations, trained Chadians, and the government would elevate the country's market economy.

Most governments have complained of having no national funds. Some have shown little interest in financing projects. However, they can borrow funds from local banks and pay back the loans with proceeds from the projects. Additionally, each African country has a communication infrastructure; telephone network and airport technology. This means any country can participate in the development and sustenance of information networks and low cost communication technologies that support national development efforts. Chapter Seven identifies such technologies and describes their usefulness. Lastly, it examines different forms of information technology necessary for development in Third World countries and expands ideas regarding the installation of national information centers.

Chapter Seven

Using Information Technology for Grassroots Development

Overview

This chapter examines the last phase of the traditional micro economics theory reflected in this book. It assumes that an information industry undertakes the production, distribution, and preservation of information for use. The entire process requires money, expertise, and knowledge (information). Economists have used money to analyze and compare markets. Apart from serving as a medium of exchange, money is a measurer and preserver of value. It is used to evaluate markets. Similarly, information is a component of any market. It is a medium through which decisions are reached. Since information regulates the value of money and vice versa, money can be considered a distributor of information. Additionally, the value of a product or commodity depends on the amount of information available to the prospect about the product or information provided by the creator of the commodity. Its creator determines the price of the product. Whether it is bargainable or fixed, price informs a prospect about a product's value. Money can also measure the value of information and information can be measured using money. Today, more knowledge is acquired through purchased equipment; telephones, computers (internet), television sets, newspapers, and other information technologies. The more equipment one has the more access one will have to information. Information or knowledge is also seen in this chapter not only as an end product, but also as a production agent. Knowledge "informs, " that is, it "forms within"—generates ideas. Knowledge will be examined as a commodity—an agent of development for Third World residents.

The chapter further examines the economic value of information. Itposits that information is a commodity, which can be used to exchange or improve services and situations, if distributed properly. It also examines information efficiency in the improvement of natural situations. Here, efficiency functions at four concomitant levels: (1) the product or service should be useful to the consumer and should visibly improve to the consumer's living standard; (2) the product (telecommunication equipment) or service should be rendered at the lowest affordable cost; and (3) the product and service should be dependable and sustainable.

That utilitarian principle, using products and services to satisfy wants and needs, is highly applicable to Third World countries that are to utilize their resources in order to influence inevitable changes in world economy and information volume. Along this line, policy makers involved with Third World countries, and forces that control world economies and information industries, should see information as having ethical and moral qualities, and hence they should reorganize Third World governments toward using information for the augmentation of their economies.

Although developing countries are not financially equipped to obtain new information technology, they have realized the importance of using infotechnology in facilitating human life and in supporting local development efforts by leading the privatization of the telecommunication industry (see Chapters Three and Six, respectively). Four major premises have been identified in this chapter for understanding the relationship between information technology and grassroots development; information acquisition systems, biodiversity conservation, education, and local incentive.

Information Acquisition Techniques

Generally, the telecommunication industry in Africa has made some progress in using IT to provide mass audiences with general information. Eight percent of all television stations are equipped with regular movable video cameras, monitors, and video recording equipment, as opposed to the early 1970s when there were no local television stations. However, most of the news anchors are still using cards for prompting. Their use of satellite and related digital equipment is limited to special events like sports finals, visits of high profile personalities, and national events.

While the audio visual media sector is not well equipped, the newspaper, the commonest source of public information, has itself not improved dramatically over the last decade, despite the advent of IT, due in part to government censorship and limited funding. Audio cassettes and recorders, manual cameras, microphones, pens and notepads are mostly used for the collection of news and information. Only in the late 1980s did reporters become familiar with the use of laptops for the transmission of data from the field to the newsroom. Journalists accompanying government officials abroad, especially to industrialized countries, became acquainted with Pentium and Macintosh computers and other multi-media systems for videographing, faxing, audio-recording and e-mailing texts back to their national stations for immediate broadcasting. The absence of this state-of-the-art equipment in most African news rooms and news agencies certainly retards their ability to report accurately and promptly. Since most local media cover soft news, laden with commentary, biased criticism and footage collected in the field are usually awaited before printing/broadcasting can take place, the susceptibility of African news consumers to reporters' personal opinions is compelling. Moreover, since most newsrooms are not equipped with computers and satellite systems, reporters, unlike Internet users are often unaware of major world events, unless they have access to telex which, only selects news items. Hence, most stories published on international subjects do not have significant news value.

Computers and Internet services are costly for African telecommunication ministries to sponsor because of their meager annual budgets. However, governments can increase their telecommunication budget to accommodate a high exploitation of information. As African nations are opening their doors to democracy, it is incumbent upon them to provide more funds for information equipment and services, in order to enhance the people's right to information. The more people know, the more enlightened they become. With enlightenment comes tolerance, understanding, and discipline. Governments and private media businesses that cannot afford computers should purchase beepers and fax machines for their respective reporters who will be able to contact head offices and meet deadlines. The quality of newspaper print should not be ignored; legible words have higher market value than ink-ridden pages. Many readers have refused to buy newspapers because of ink-blurred words, even when the story had high value. The media business manager should realize that investing in new affordable information technology can enhance business.

While the industry has been slow in investing in infotechnology, other sectors of society have realized that commerce can be carried out effectively only through electronic networks with the capacity of transmitting and storing large volumes of data. In Zambia, nine rural districts had network access, which provided a high level of staff morale and performance. CD-ROM, World Wide Web, video-conferencing, e-mail, and other interactive media have increased learning resourcesbetween Nairobi, Capetown, Paris, London, New York and other cities. These distance learning facilitators have helped people worldwide to share useful information about each other's communities. In such areas in Africa, workers felt important dealing with foreign electronic equipment and personnel. To be in contact with, or see colleagues in, a live classroom session abroad is a privilege; not many people have had that opportunity. The chance to interact freely with a foreign entity is a novelty because information about the international community normally has reached them through government censored press.

Similarly, the convergence of telecommunications computers, satellites and fiber optic technologies—has expedited the spread of information and has sustained a rise in the cost of processing, storing, retrieving, and transmitting information.

The next segment examines the capacity for different non-traditional media agents to contribute to physical capacity building and to the manifestation of human capital.

Biodiversity Conservation

There has been an increase in environmental degradation in Africa since the end of the Cold War, and that has stalled technological advancement. With no threat to their foreign policies, industrialized countries have been using Third World fauna as dumping grounds for industrial waste products which have caused chronic illnesses and deaths. In addition to dumping waste products, the rate of recurrent drought, deforestation, and waterborne diseases across the continent has increased. According to *Findings,* a newsletter published under the auspices of the World Bank (November, 1996), 273.5 million rural residents and 45.6 million urban residents do not have safe drinkable water. By the year 2000, about 350 million Africans will risk living in a water-scarce environment, as deforestation continues. This problem needs to be addressed through education and technical assistance.

Other areas of biodiversity conservation which deserve attention are household water security, food security, water quality, human health, and wet land protection. Although West African countries have enough water on a per capita basis, high population growth and investment on new infrastructure, irrigation systems, hydro electricity and other economic activities create an additional demand for water supply. About 320 million hectares of vegetated lands have been degraded, causing flooding, reduced ground water recharge, and reduced stream baseflow. These new activities only support the need to educate residents on how to sanitize water and provide clean water for food production and consumption.

The decline in food production and lack of food security in parts of Africa, particularly the rural areas is directly linked to the loss of biodiversity. Food production can be increased by augmenting food productivity per hectare, clearing new turf (forests, bushes, and swamps), and by storing information on farming, fishing, and food preservation methods on audio and video systems. Today's farmers are switching to mechanized techniques and forgetting ancient farming and food preservation techniques. Conversely, African ancestors lived long because they used sophisticated information techniques to cultivate and preserve food from the fauna and flora. Messages were passed on from one generation to another and people mastered farming methods through instruction, observation, and practice. During the planting and harvesting seasons, farmers erected scarecrows to ward off intruders, in so doing they were able to preserve crops. Thus, scarecrows served as information carriers.

Because information is an important element for supporting development activities, examining the characteristics of the equipment used in preparing it, and the services provided is equally important. Hence, the role of the informer is as critical as the information itself. An informant, person or equipment, is the message itself because the entity has the ability to reshape the content, structure, and meaning of the message before delivering it. Therefore, an informant can control an informee's way of thinking and acting. Expatriates are major informants, —carriers of information—on technical issues in Africa.

The Case of the Expatriate

Expatriate service can be seen as a paradigm for assessing the exportation of communication technology because an expatriate is a bona fide technician who brings into a country a pattern of behavior and a set of skills uncommon to local residents. The process of utilizing skills to train people and/or work individually to produce results involves interpersonal interaction. Hence, for an expatriate to train people, he/she needs to learn their habits. Africans have experienced an influx of expatriates and development workers with minimal visible improvements in African's living standards. These expatriates enjoy a lifestyle that most Africans do not afford. For Africans who wonder whether the expatriates are there to assist them or to elevate their own social status, such practices are taboo. In the late 1980s when Tanzania, Nigeria, and Cameroon were experiencing economic crises, it was normal to wait in line for several hours in the supermarket, except for expatriates who were served first in exchange for tips. This nonverbal communication created conflicts between residents and expatriates and deterred the much-needed interaction among local groups and between them and the expatriates. However, expatriates help to expand local economies by buying locally made products, leasing real estate, hiring local residents, patronizing public transportation, and marrying local residents.

The difficulty in dealing with target groups has also been caused by limited knowledge of the indigenous culture. What development practitioners learn in classrooms, may be relevant but inadequate for use in the field. When sent to the field to gather information, or assess community needs, they find the social level of the people rather complex. For instance, in a program projected to last five years, much of the time is spent on logistics, rather on productive work. During the first and second years of the program, expatriates study the landscape and its residents. By the third year, they begin to understand the modus operandi of the country. During the fourth year, they begin the operation of the program, and in the final year, they rush to complete the assignment in order to leave. In essence, much of the time is spent in the field learning about the environment, rather than evaluating the progress of the program for future quality research or improvement. Expatriates appear qualified to manage projects or serve as consultants to companies seeking business opportunities after receiving brief training and having made several trips to Africa. As a result, community members whose cultures they never fully understand provide them with half-baked answers that become the bases for their evaluation.

Another dilemma is the presence of multiple cultures. Since different cultures are practiced in different enclaves, it is not easy to generate a lucrative telecommunication market in Africa. The complexity of each nation requires specific research and answers. This means survey results in one country cannot be applied to problems in another country. On the contrary, information sharing among natives can lead to efficient and cost effective environmental management. For expatriates to communicate effectively with natives and telecommunication officials, their level of consciousness about Africa must be raised. To begin the process, foreign telecommunication companies and their employees must listen, take instructions from African experts and allow the latter to plan and present their agendas during international debates. In addition, they must be trained to understand and work with urban Africans, who practice EuroAfrican culture. That category constitutes qualifying prospects—the most available market for foreign products.

Experts communicate good intentions and ideas on how to perform environmental and conservation work but operate within organizational bureaucracies that communicate their own agendas. Similarly, development agencies operate on a territorial level by relying on their associates for information but are eventually misinformed on which development factors to invest in, since the former (experts) do not fully understand the target group's culture. So experts must closely examine the context in which Africans impact on the environment prior to providing assistance them any form of assistance. Because of the existence of multiple cultures and complex ethnic composition, Africa's problems are too complex for simplistic and generalized solutions. Thus, as bona fide investigators and investors in the local economies, experts have an obligation to research beyond superficial observations. Certain methods of managing the environment may be ecologically adaptive and useful only in a particular environment. For example, a management strategy may be adaptive to specific ecological circumstances because of soil type or terrain. However, the same management technique may not be feasible in another country, because of political conditions. The expert must be aware of these shortcomings.

Experts should first visit rural communities and interact more closely with local people, in order to understand specific habits of urban residents who often nurture a naive view of international products and providers. Grassroots interaction may be costly and time consuming, but it is the most feasible approach to identifying people's needs and improving services. Understanding Africa and disseminating relevant information for development and conservation also require obtaining information from "non-traditional sources." The most relevant information may be from sources totally unaware of the existence of infotechnology. A grandmother who never received a formal education and who was a farmer has invaluable knowledge about agriculture and how it is related to her immediate surrounding than an expatriate whose knowledge of agriculture is primarily drawn from books and theories, rather than from practical experience. The grandmother is an effective information resource. She may provide insights on her farming methods which agro-journalists and biodiversity communication experts could use in developing materials for telecommunication providers and information technicians. The major premise here is, infotechnology should not be dumped on the masses, rather they should participate in information production and influence the creation of appropriate technology in order to expedite grassroots development.

The Non-governmental Organization (NGO) Dilemma

There is an underlying assumption that the proliferation of NGO creates a barrier in the exchange of information on biodiversity at the local level. NGOs, like other development organizations, do not always act on what they know. Relinquishing some aspects of their control in the development process will require giving up some of the market. The institutional framework of African NGOs has major problems. NGOs operate in a fragmentary fashion and lack consensus and cohesion, not only because of the influences of the universities from which they received training, but also because they have virtually no input into national politics. Experts trained in a French institution may have a different concept of development from those trained in an American university. Their dilemma with the governmental system is that it applies the top-down approach to development. Moreover, international development agencies sponsoring projects in developing countries have been interacting directly with government officials by signing contracts and planning projects, without considering the input of NGOs who understand groups at which projects are targeted.

African NGOs realize the need for capacity building in local institutions. However, the inconstancy of funding helps to cripple their efforts. When funded projects run out, projects are discontinued. Local NGOs like Opportunity Industrial Center (OIC), have suggested setting up institutional programs, as opposed to five-year contracts, in order to create their own funding and become self-sustaining. However, decision-makers in some funding institutions and high-level workers have striven to maintain the status quo—keep NGO personnel in business—in order to save their own jobs. If NGOs become self-sustaining, contracting or funding agencies will be out of business. Thus, in order to establish a self-sustaining community, effective telecommunication services should be provided.

Framework for Providing Telecommunication Services to Targeted Constituencies

Urban Communities

African urban communities are a market for biological resources. People in urban areas use local commodities such as medicinal plants, game meat, and fiber in designing handicrafts. Generally, there is an entrepreneur who purchases these products from rural market and sells them to buyers from cities. It is crucial for communication practitioners to develop and present preventive materials during town meetings, in order to educate the residents on the disadvantages of destroying such ecological structures. Unless residents see video tapes, slides, or live graphic images of rare animals being slaughtered to make shoes, or young plants being eroded and causing drought and famine, they will continue to undermine biodiversity conservation. Monthly town meetings and video displays are therefore necessary to help residents in conserving their ecosystems. That approach to disseminating information in urban areas would curb the high demand for biological resources. However, understanding the impact of the loss of biological resources alone will not necessarily change residents' actions. They need the practical alternatives.

To understand the impact of alternatives, one should examine the consumption of biological resources in urban communities along gender lines. Female entrepreneurs in such communities are influential in determining the future of biodiversity conservation, because they control the use of as wood and charcoal consumption. Poor women use charcoal and firewood in lieu of electric or gas cookers. A bunch of firewood extracted from a tree trunk weighing 70 pounds is used about twice to produce charcoal for cooking and ironing, whereas the same amount of gas could be used for months. Moreover, felled trees have the capacity to fertilize the soil for agropastoral benefits. By using trees to generate heat, a short-term benefit, the consumer diminishes the import of that resource. Wood produces smoke which through prolonged use, destroys the lungs and causes cancer and other kinds of chronic illnesses. Even urban residents also extract wood from forests to generate energy, despite the availability of alternative affordable resources like gas and electricity.

In order for people to set new standards on charcoal consumption, they should receive relevant information on the negative impact of charcoal consumption on the economy and the environment. Additionally, telecommunication industries should produce new equipment that can be used in educating biodiversity destroyers. This equipment should be able to supplement the use of videotapes and live satellite programs and should inspire women to participate in researching alternative energy sources. If women are excluded from the campaign, they may not be willing to accept the new technology.

Rural Communities

Rural residents living in areas with the largest concentration of biodiversity have unrecorded traditional knowledge of biodiversity conservation. The conservation community, agro-journalists, telecommunication providers, and development communication practitioners must seek current indigenous knowledge about techniques of conserving biodiversity in order to prepare appropriate campaign materials that would foster local initiatives. They must understand that altering the priorities of rural people can affect the utilization of biological resources. Situations that appear to be in the agrarian community's best interest may not necessarily be politically neutral. For instance, the government extracts rural resources to finance industrialization and appease the urban elite but pays low prices to farmers for their crops in order to save money and construct hi-tech roads and buildings in urban areas. That practice has generated skepticism and mistrust among rural communities.

Students/Youth Education

Theoretically, this group receives formal education in order to improve social living standards. The young educated African is caught in crossroads of development. Prior to colonialism, children received practical education from their parents which was designed to serve the individual, the family and the community. The knowledge they now acquire in the classroom has a significant impact on their environment. Yet, most of their education is based on a European curriculum that has limited bearing on their immediate needs. Even students pursuing agro-pastoral studies are taught mechanized techniques, whereas their communities do not have the capital or resource to implement the students' ideas. Such education underprepares them for the challenges of conserving biological resources that are the bases for their survival.

Modern African education has been adapted from a colonial legacy of self first (self-reliance). An educated African with no knowledge of computers and no exposure to foreign media has limited chances of progressing in modern society. Hence, African educators should reevaluate their academic curricula in order to tailor knowledge to fit Africa's development goals. Western education should enhance, rather than hamper, the objectives. The youth in particular should become more active in pursuing biodiversity conservation since they currently form a growing majority of the continent's population. They should be provided information on the conservation of their environment. In American schools, students do science projects which helps them to exercise their creative intellect. Youth in Africa also should be challenged through locall-sponsored media campaigns to research ways of resolving environmental problems. In addition, government officials should implement programs that allow young energetic people to learn crises management techniques. Although they do not have technical and financial resources to undertake the task, young people have creative incentives to design information technologies and plan strategies of spreading information on biodiversity conservation, especially in the villages where they are often admired and depended upon for leadership. Moreover, some of the villagers are potential farmers and poachers. Hence, their participation in environmental protection cannot be ignored.

Youths and communication experts can use youth centers to produce music and drama on biodiversity conservation and other development factors.

Teachers

Educators have the opportunity of conveying information on biotechnical matters to the general public. Since teachers are respected in the community for their ability to nurture minds and transmit information to large groups, it is imperative that communication trainers use teachers to spread development information. In order to obtain successful results, teachers should be trained on how to use technology in collecting and disseminating information.

Governments

This category is responsible for making and enforcing policies. Due to a high level of bureaucracy, there may be difficulties in exchanging information on environmental protection, so the government should use communication experts to educate the masses. All ministries, especially health, wildlife, environment, education, agriculture, forestry, health cultural affairs and trade should interact with each other to provide useful information. Governments should help the poor, especially women, who dominate Africa's population numerically, gain access to information on land use, credit, and extension services.

Applying Relevant Communication Technology

Technology is a double-edged sword because it can either create or destroy. For technology to have greater use, it has to be locally adaptable. If adaptation does not take place, technology has not been transferred. Transference presupposes utilizing products and services brought from a different culture. Technology can only help in improving production if properly applied. The following section briefly describes the role of low cost information technology in channeling development messages.

i. Transition/Mobile Radio Sets

Many development studies show that radio is the most cost-effective and affordable tool for disseminating development information in both rural and urban areas of Africa. Most people, including the rural poor, can afford and depend on radio broadcasts extensively for useful information for the following reasons:

1. Programs are broadcast almost throughout the day;
2. Programs are broadcast in local language;
3. Radio technicians often interact with the rural populace when recording rural-oriented programs;
4. Government-owned radio stations that make up 99 percent of the radio service in Africa have powerful modulation frequencies that can reach people approximately 200 miles apart;
5. Although it is a one-way form of communication, radio becomes interactive when people listen and discuss issues. It is important to have an organization that designs and processes development information for broadcast.

ii. Television, VCR, and Other Recording Devices

Visual communication is obviously more effective than radio in spreading development information. However, television sets are scarce because of their cost. Since the urban affluent are more likely to own a television set or benefit from television programs, specific programs should be targeted at that audience. With an average of 50 television sets being shipped per month to over 700 million viewers across national borders in the 1990s, especially in sub-Saharan regions, as opposed to 300 viewers per television set in the 1980s, one would expect heavy exposure to broadcast information. Following a dramatic increase in adolescent population and lower middle class workers, television purchasing and viewing rates have increased. While more Africans are becoming exposed to television content, fewer Africans are using VCRs, because of a drop in the purchase of foreign videotapes. Hawkers smuggle them across borders to evade tariffs. Also, corrupt immigration and port authorities receive bribe from importers and do not declare the product. Thus, it is difficult to determine the accurate number of television sets and VCRs in Africa or quantify their impact in influencing viewer behavior.

Japan, Germany, France, and the U.S. are primary exporters of television sets, to Africa. Meanwhile, television sets assembled in Nigeria are available in more West African homes than those imported from Japan, the U.S. and Europe because they are cheaper and adaptable to Third World conditions. Hawkers smuggle them across borders to evade tariffs. Also, corrupt immigration and port authorities receive bribe from other importers and do not declare the product. Thus, it is difficult to determine the accurate number of television and VCRs in Africa.

Due to increasing economic crises ravaging Africa in the 1990s, the importation of Western TVs and VCRs for sale has declined 1998, although interest in using the technology has increased by about 70 percent , due to Africa's continued exposure to Western culture via magazines and other media. Those exposed to European culture through school, church, nightclubs, city, or government-managed offices have also cultivated a dire interest in Eurocentric values laden with Marxist ideas of self-enrichment. The acculturated feel superior to villagers because they can communicate verbally with major law-makers and law enforcement officials, in a European language (Ngwainmbi, 1995a) which the villagers do not understand. Through such acculturation, the "educated" (those exposed to European ideas) have maximized an interest in foreign entertainment technology. While local media specialists broadcast local programs for national audiences, VCR owners prefer using imported videocassettes.

Thus, in order to enhance local culture, more local programs should be broadcast in local media stations. Unless residents fully understand their own culture, they may not be able to exploit a foreign culture to their advantage. An African videographer has more ideas when shooting a video for local audiences than a foreign counterpart. While no African has yet invented a video camera or television set, the African videographer will know what scenery can benefit the local viewer because he understands the values of his people. African videographers do not need expatriates to show them what to shoot, although these local experts require adequate training from manufacturers prior to utilizing such products. It is also important for foreign manufacturers and local experts to discuss communication technology that is applicable to African culture prior to manufacturing and exporting it.

Manufacturers risk losing profits by tailoring their products for relatively unidentified markets like rural Africa. Moreover, the increasing number of field managers, information technicians and video camera purchasers in Africa reinforces the need for foreign video manufacturing companies to design and sell African-oriented products at affordable costs to local farmers, small business groups, schools and health clinics. Like video recorders, tape recorders have proven to be effective for information specialists and field managers in the processing of data on rural and urban populations. The need to videotape marketable products and to present them to sponsors for specific assistance cannot be underestimated. If field managers and media technicians use recording equipment to capture live pictures of deforestation and related biodiversity practices, as well as images of fertile land, not only would the agrarian continent become aware of the long-term economic value of their environment. They would develop an interest in obtaining and preserving videos for future generations. If more African-based videos become available in stores and exhibits abroad, there will be more global interest in African affairs, and Africa's capacity to participate in the global information market will increase.

Conversely, if Africans continue to deforest the land in search of food, farmland, and construction sites, and if industries continue to deposit toxic waste in the rivers, Africans will continue to lose sustainable water and land resources. These problems have indeed caused famine, and dependency on foreign grain, and have sparked political tensions between government and its people.

iii. Participatory Video

Although participatory video is an effective tool in communicating information, it is a new phenomenon in Africa; therefore, it must be introduced with caution. Also, because of its complex operational nature, people might become more interested in the equipment than in the message it contains. The success of participatory video cannot be predetermined; success will depend on the responses of each community.

iv. Environmental marketing

Social marketing often focuses on different behaviors of prospects and increases benefits and reduces barriers, guarantees better service and communicates persuasive messages. It is an effective way of reaching people, especially in closed societies where people are resistant to new ideas. It is a form of environmental education that can be presented formally or informally. Formal education can benefit students enrolled in environmental programs while the informal mode can be useful to uneducated and employed youth.

iv. Environmental Training Centers

The primary function of environmental centers is, they must serve as a clearinghouse for development information. Environmental centers in urban and rural communities is an effective medium for disseminating development information, and for encouraging youth to actively seek solutions to problems in their immediate surroundings.These centers can be used by project managers, development organizations, and others to give people information collectively, and to receive information from them. Information would also be exchanged through lectures, workshops, and seminars. Hence, centers should be open to all interested people. Like environmental training centers, workshops and seminars could enlighten development agencies, NGOs, trainers, researchers, and urban residents. Setting up a workshop forces residents to participate in information sharing without prior conference with their leaders. Moreover, people in rural areas do not like to be talked at, or forced to give their opinion. They believe opinions on public matters should be given by those in authority.

V. Exchange Programs

Exchange programs can be advantageous to each of the communities involved in urban areas when project managers and technicians visit project sites, acquire information, and observe the way others address development problems. Although transportation from one African country to another may be costly, it is necessary for telecommunication providers to sponsor such trips for project managers and technicians, in order to understand the peoples' needs and find solutions to the marketing of their products. For instance, fax machines may be a more profitable investment for the Ghanaian market than video equipment. In Accra and Legon where more educated and trained people live, telephones and fax machines would be in higher demand than cameras, because the former can generate more business transactions. Thus, providers should provide transportation facilities for marketing consultants to find lucrative markets in different African townships.

In rural areas, villagers can learn from each other through exchange programs. Development ideas from rural and urban residents can be channeled to both entities through public relations experts. However, this practice can only be effective if both communities put aside ethical differences and think collectively; develop ideas on how to improve agricultural production, how to conserve biological resources, and increase the production and marketing of business commodities.

Vi. Religious Organizations

Religious organization cans draw attention to issues. As trusted Godly people in spiritual areas like rural communities, religious groups and leaders are more likely to be listened to than any other group. Telecommunication providers can target such entities to promote their development-oriented products. But the performance of these groups will also depend on the political situation of each country. Telecommunication providers should not use religious entities in countries that prevent freedom of religious expression.

Vii. Village Action Groups

In African villages, action groups provide services to the community when need arises. These groups have constructed roads, laid pipelines, and assisted widows in their farms. Group members include self-motivated men, women, and youth, as well as household heads and teachers. This caste who ordinary community members respect for their leadership skills can work with communication practitioners and telecommunication providers to elevate understanding of the people's needs. Project managers should work more frequently with action groups which best understand community concerns, in order to disseminate development information. The disadvantages of using village action groups are, they may be overburdened and taken for granted. Traditionally, action groups have provided free services to both community and aid donor. For them to promote products or carry out any profit-oriented services, action groups should be compensated. Compensation itself would be an incentive to the community; it would attract the attention of the ever-increasing number of unemployed youth and would eventually sensitize them participate in respective development activities.

Viii. Computers

Computers are best suited for elite in urban communities. Government, NGOS, project managers and academic institutions are primary markets for computer use. Due to limited availability of telephone fines, computer use will still be limited. However, networks like Fidonet have been designed to overcome the limitations caused by inadequate phone lines. Fidonet provides a low-cost method of linking computer bulletin boards with ordinary phone lines. Hence, educated elite should take advantage of this facility and share development ideas.

Ix. Project Managers

The role of project managers should be contextually defined in order to facilitate the exchange of information constituencies. Project managers and other field workers can evaluate communication processes among targeted groups by observing the latter tackle biodiversity and other socioeconomic activities. They can also record the amount of knowledge that target groups have, the degree to which different constituencies share information, when and how project managers can determine the degree of its sustenance. Information derived from that process should be made known to members of other target groups. This is one way of encouraging networking among constituencies.

Summary

Through the four premises identified in this chapter, authorities can see telecommunications as a value-oriented system capable of providing market information at an affordable cost to all communities. Over 840 million people in developing countries risk dying from famine, malnutrition, and drought at the turn of the century, due to limited information available, deforestation, and other ecological problems. Telecommunication providers can alleviate such conditions by manufacturing more affordable and easily applicable information technology for the poor, who numerically dominate the population of developing countries. The training of personnel that can fully comprehend the conditions and values of local residents, their knowledge of new technology, and a competent workforce is imperative. Also necessary in bringing economic independence to Third World countries and in stabilizing their participation in global endeavors is a redirected community spirit.

Recommendations

Information empowers and liberates, therefore an information system should be set up to help African people realize their development objectives. If the following conditions are met, Africa could be among the foremost beneficiaries of the global info-economic markets by the year 2025.

1. Telecommunication officials should regulate the obsolete telecommunication industry by allowing foreign companies to win contracts for the installation and collaborative management of information systems.
2. There should be an improvement in the creation, processing, and sharing of development information that would help small-scale businesses improve quality of products and marketing techniques.
3. Promote privatization and encourage competition among local retailers and providers. That could create more jobs.
4. A telecommunication network should be installed in government offices and research centers to expedite the storing, retrieval, sending, and sharing of general development information, and for the dissemination of information on government-funded programs (see Charts 7.0 and 7.1).

That process could advance the creation and exchange of quality data and eventually improve the exploitation of national natural resources (e.g., minerals and human capacity). With a significant rise in literacy rate and work-fit population, a nation-wide computer education campaign should be easily carried out simultaneously. Youngsters and civil servants should be trained to use the computer in order to improve work output, and compete for jobs and business opportunities elsewhere. That would reduce the degree of idleness, crime, poverty, and indiscriminate recruitment into the army, while advancing interpersonal and intertribal understanding.

5. Local officials can indeed expose institutions to the advantages of installing new telecommunications infrastructures. For example, following the installation of a fidonet in Djamena in 1995, Chadian officials along with the French government attempted to obtain voice systems in order to establish effective message transmission on the internet.
6. The Ministries of Education and Cultural Affairs should in-clude environmental and videographing courses in school curricula, in

Chart 7.0: Internet Education: Connectivity Flow for Government-funded Programs

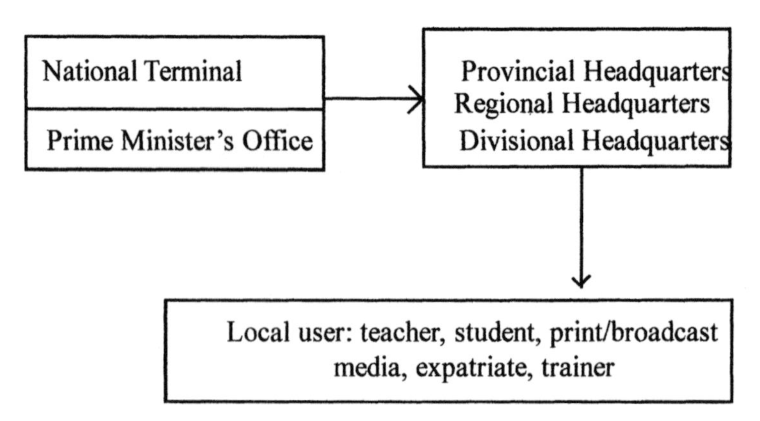

legend: Arrows show route to be followed by local server

Chart 7.1: *Message Distribution for NGO-Sponsored programs*

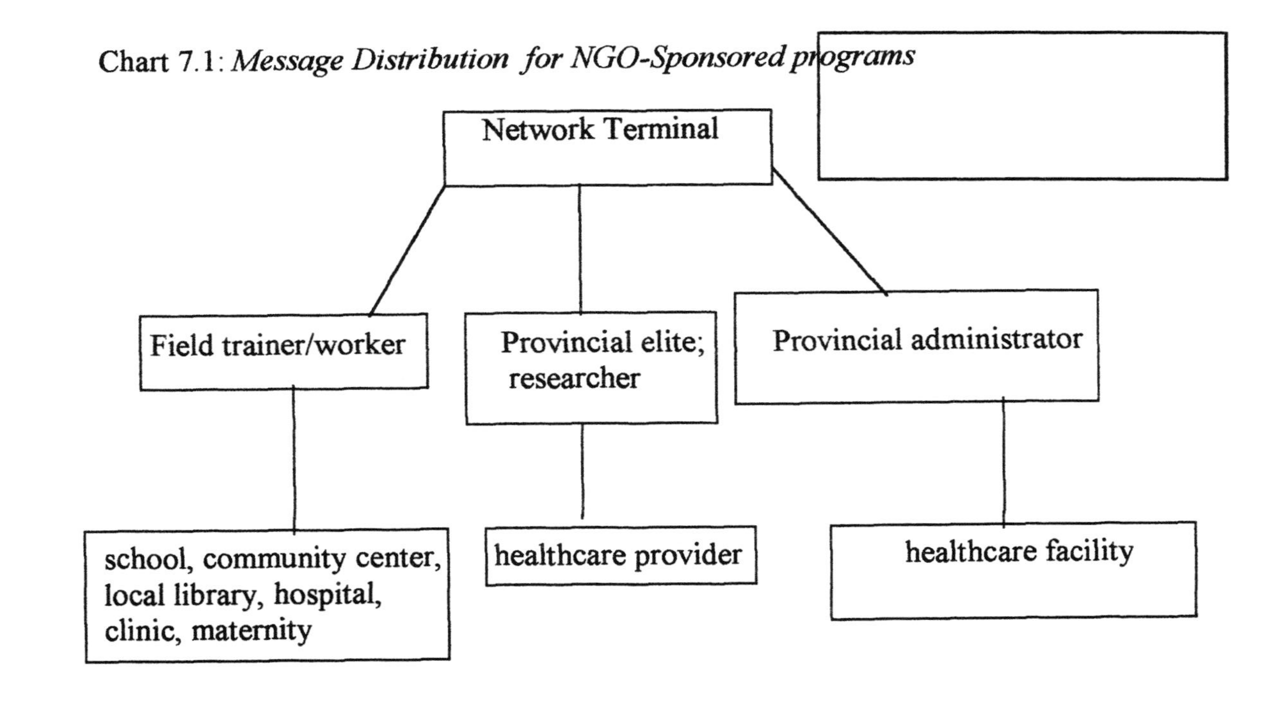

> order to facilitate the efforts of international development agencies. Environmental education should center on issues that affect the immediate surrounding. People should be taught how to utilize such resources to generate personal income. This would help reduce the practice of destroying biodiversity.

Landlocked countries have an opportunity to acquire and connect a powerful network with other regions. UNZANET's connectivity in Uganda, Tanzania, Zimbabwe, and Malawi and with international stations clearly indicates not only the progress made by officials in those countries toward promoting a global information economy, it exemplifies the potential for landlocked countries to communicate—improve the exchange of market, climactic and social information. A regional internet system could open up agriculture markets and help servers in neighboring countries process and share messages on effective farming techniques, better healthcare practices, as well as on weather and consumer behavior.

Pilot studies on how electronically processed information can be used to improve business -to-business and business-client transactions are required. For example, an Internet project could help business entities share ideas on efficient farming strategies in all regions, learn about the world, transact inter-office business and develop incentives for self-reliant growth. That should consist of meetings between officials in the ministries of culture, information, economic planning, local business owners, and information experts. Also, computer training of these entities should take place. The study should be done in three stages: a need assessment, or fact-finding mission, installation of the technology, and periodic evaluations of the project for improvement.

African officials in Washington, D.C. embassies and in Africa should set up meetings with potential providers. The process should include:

1. Hiring computer hardware and content consultants to survey the landscape and make estimates for the installation of suitable computers and software packages;
2. Holding meetings with information technicians and local officials to discuss cost and manpower (kinds of services

needed). The discussion should include hiring computer technicians (help-desk associates), network administrators, etc.

3. Discussions regarding the content of information for the Internet. Policies regarding the inclusion or exclusion of Internet information must consider cultural values.

This book strongly recommends the establishment of an information telecenter in each country to be linked to the global information supermarket. As stated in the previous chapter, local information centers can become outposts for national development activities if managed by independent organizations. The centers can be locally owned and operated by micro businesses--a franchise which provides low cost services to residents in rural/poor areas. NGOs, civil society groups and local governments can also play a role in the initial stages of the project. Because of NGO expertise, the political influenceo of local organizations, and the judicial privileges of governments, foreign sponsors must collaborate with such entities to successfully sponsor the project. Services should be provided based on the most pressing needs of residents. Immediate services may include telephone and voice mail to enable residents reach their family members, fellow tribesmen, friends, and business associates in big cities and even abroad.

Thereafter, computer services should be provided. Services should include training at initial, intermediate, and advanced levels, based on trainee's level of education. Training should include learning common languages used by most people (English, French, and Spanish) since 98 percent of rural residents only know their own native language. Also, since such potential users come from very poor backgrounds, free training should be provided to all. Conversely, income earners and striving businesses make up less than 10 percent of the rural population. So charges for training and property rental may drastically reduce the number of interested persons. Once training has been completed, trainees can use e-mail, internet and information storage and retrieval services to find jobs and obtain information that can help them alleviate famine, malnutrition, poverty, disease, and other setbacks to personal development.

At the regional level, information centers should be set up to provide technical information for the smooth running of the country. As most government workers serve in towns and cities, there should be a center that provides educational, effective institutional management, and business development (including cooperative and agriculture) ideas. A national information center should have an international website to help users in developing and developed countries surf social, economic, and cultural messages. The center should also serve mostly diplomats, and government officials while residents should have access only to certain websites and software in order not to be exposed to material that can upset national security.

The capital costs for installing information equipment and its personnel ranges between $20,000 and $30,000 for rural center, $50,000 for a regional center and about $90,000 for a national center. However, only about $2,000 may be spent in maintaining equipment donated to the rural area per year. If it is leased, capital requirements may range between $12,000 and $35,000 for all three centers.

Based on its values, each country should develop a website from a template. The following segment (Chart 7.0) describes the sort of information to be found in a national website.

Surfing Bulletins

News Updates

This window should contain information for those logging on the website in search of new material; upcoming events, recent changes in government, new currency exchange rates, new available services, new decrees affecting foreign entities, tourism, travel, weather, etc.

Culture

This section should focus on recording, storing, and making accessible the history and culture of each clan/tribe to browsers. Written folklore, audio and video streaming copies of songs (panegyrics), storytelling, and traditional dances should be available here. Also, published and on-going research on a given topic should be displayed.

Chart 7.0: Proposed Country Website on the Internet (www.country.com)

Government Bulletin	Tourism Soccer	Embassy	
Culture	Conferences/Trade Shows	Post Office	PSA
Current Events	Public Services	News Update	Weather
Professional Associations		Professional Abroad	Chat Club
	local marketplace		

Public Services

tion here. This site might become the marketplace for everyone and might help bolster the local economy if users display their addresses and information on expertise.

Sports/Soccer

Player profile, scores of First Division matches, and the national team's participation in international competition should be updated here. Sports agents abroad will find this site helpful in selecting athletes for professional careers in Europe and elsewhere. For the country's citizens abroad, this will be an appropriate bulletin to get tidbits which are not usually covered in sports pages in their community.

Current Events

Video clips of news from various sources as well as other news making headlines among international media should be displayed here. Also, the visits of VIPs should be updated.

Post Office

This window should serve as a directory for people and organizations seeking mutual interests. Email, postal addresses, phone numbers, and names should be placed in alphabetical order to facilitate the search.

Local Market ("Uwe"-market in Kom Language)

This should be the American version of a phone book. Low scale

Local Market ("Uwe"-market in Kom Language)

This should be the American version of a phone book. Low scale businesses, firms and manufacturers can display their commodities and services here. This would be an opportunity for unknown talents and products to reach international markets.

Conferences and Trade Shows

A schedule of upcoming social, political, and economic conferences and trade shows anywhere in the country should be displayed here. The site should also include conferences and trade shows taking place abroad that involve the country. Contact information (phone numbers, email, postal addresses) and places capable of hosting meetings, trade shows, and conferences should be found here.

Chat Club

This section should house public discussions, dialogue, electronic conferences, news teams, and video or voice conferences over the internet. Browsers will be able to dialogue with others nationally and internationally.

Government Bulletin

Here ongoing discussions between important visitors, personalities and organizations on specific issues should be found. Professionals in and outside the country should use this forum to chat with colleagues locally and internationally. Another focus should be the provision of information on services that improve national investment, trade and diplomatic relations between the country and the significant other.

Tourism

Through this site, browsers can find information on maps and travel facilities. Increasing use of this site can help build the local economy when tourists make trips to the local area. Information should focus on the nature, beauty and value of local culture. It should also have images of different tourist sites, information on the distance to such sites, on transportation facilities, and road conditions.

Professionals Abroad

This should be a directory of professionals in the Diaspora. The site will be useful to major international organizations like the World Bank, Inter-American Development Bank, etc., in finding consultants for projects to be done in a specific country. The directory should contain the consultant's name, address, qualifications, experience, and a list of his/her references.

Embassy

This section should have a list of a country's embassies, their staff members, and responsibilities to help users establish contacts. It should also have visa and country-specific business data to help people coordinate their travel plans and fulfil immigration requirements accordingly. It could increase the value of dealing with legal issues and efforts to expand business opportunities.

Weather

This window should display information on the daily weather of major cities and towns and advise visitors on what to wear. It could also help farmers determine when to plant crops or carry out different agricultural and social activities.

PSAs

This will be the right place for retrieving public service announcements. Issues that require government or public intervention; finding the right place to obtain assistance, locating a missing person, knowing the location of a relative (especially those abroad looking for a long-lost friend or loved one) can all be documented here. For those at home and abroad trying to send or receive money, even a package, this site would be most helpful.

Professional Associations

This will be an important site for scholars, researchers, and academics seeking tips on conferences, new publications, activities of professional associations, innovative teaching techniques and instructional materials from local and international entities. The main purpose of this site would be to facilitate information sharing and caucusing.

References

Adam, L. & Hafkin, N. (1993). The Padisnet Project and the future Potential for networking in Africa. In *Electronic networking in Africa: Advancing science and technology for development*. Washington, D.C.: AAAS.

Adekola, S.A. (1989). Planning communication systems: Critical models and variables. (pp.10-23). In T. *dimensions of national communication policy*. Lagos, Nigeria: Center for Black & African Arts Civilisation.

Akwule, R. (1990). Modern Telecommunications Policy & Management Trends. In *Proceedings of Africa Telecom 90. Harare, Zimbabwe, December 3-6.*

Arewa, 0. & Adekola, N. (1988). Redundancy principles of statistical communication as applied to Yoruba talking drum. *International Zietschrift fui Volker-und Sprachenkunole Sonderadbruck, 2,* 75, pp. 185-202.

Baranshamaje, E. et als. (1995, March 29). *Increasing Internet Connectivity in Sub-Saharan Africa: Issues, Options, & World Bank Group Role*. (Draft). Africa Region, The World Bank: Washington, D.C.

Barnet, R. J. & Cavanagh, J. (1994, May/June). Creating a level playing field. In *Technology Review*. pp. 46-54.

Barrym, M. (1980). The political economy of regulation. NY: Columbia University Press.

Baptiste, A. (1994, July). World Bank information technology lending policies in Africa. Washington, D.C. : LOC650. Unpublished position paper.

Becker, J. (1994, July). *Acculturation and Technology Transfer: Telecommuni-*

cations. Paper presented at the 19th Scientific Conference of the International Association for Mass Communication Research. Seoul, Korea.

Becker, J. (1990). Aus der Gescgichte iernen? Die Zukunft des Telefons. In *Nachrichtentechnische Zeitschrift, 5.* pp. 360-367.

Brenner, C. R. & Capelo, R.G. (1987). *VCR Troubleshooting and repair guide.* IN: Howard W. Sams & Company.

Buhler, R. (198 1). Effective acquisition of hardware and software in developing nations. In J.M. Bennett, & R.E. Kalman (Eds.). *Computers in developing nations* (pp. 2 15 -2 2 1). NY: North Holland.

Cabanellas, G. (1979). The Argentine Transfer of Technology Law: An Analysis and Commentary. In *Hastings International & Comparative Law Review, 4.*

Capacity Building for Electronic Communication in Africa (1994) 8, (4), pp. 23-24.

Cappel, J. (1995). A Study of Individuals' Ethical Beliefs & Perceptions of E-mail Privacy. In *Journal of Business Ethics, 14.*

Cook, W.J. Impoco J. & Cohen, W. (1994, Jan. 17) . *U.S. News and World Report*. p. 56.

Davies, M.D. (1985). Appropriate information technology. In *International Library Review. 17*, pp. 247-258.

Davidson, R.O. (1994). Energy & Power Development in S. Africa: Challenges & Prospects for the Prospects. In *Science in Africa: Energy for development beyond 2000*. Washington, D.C. AAAS.

Davidson, R.O. (1992). Energy issues in Sub-Sahara Africa: Future directions. In *Annual Review of Energy & Environment, 17*, pp. 359-403.

DeLoughry, T. (1995, Sept. 15). Copyright in Cyberspace: New Federal Rules Proposed to Oversee Electronic Intellectual Property. In *The Chronicle of higher Education* XLII, (3), pp. A22-A28.

DiSalvo, D. (1995, Winter). Higher learning at Cybernet U. In *Campus*. pp. 18-19.

Dominick, J. Sherman, L., & Copeland, A. (1993). *Broadcasting/cable and beyond: An introduction to modern electronic media*. New York: McGraw Hill.

Dordick, H.S. (1995). Charting the Future of Communication Services. In F. Williams & J. Pavlik (Eds.). *The people's right to know: Media, democracy, & the information highway*. Hillsdale, NJ: Lawrence Erlbaum Associates, Inc.

Electronic networking in Africa: Advancing science and technologyfor development (1993). Washington, D.C.: AAAS.

Fact sheet: First annual report of the trade promotion coordinating committee (October, 1993). In U.S. Department of State Dispatch. pp. 711-713.

Fidler, R. (1995). On prospects for citizens' information services. In Frederick Williams, & John Pavlik (Eds.). The people's right to know: Media, democracy, and the information highway. Hillsdale, NJ: Lawrence Erlbaum Associates, Inc.

Gloria E. & Nurudeen A. (1994). Colonialism and African indigenous technology. In *African Technology Forum, 7*, (2). pp. 2 7-28, 46.

Gross, L. (1997). *Telecommunications: An introduction to electronic media. Madison.* WI: Brown & Benchmark Publishers.

Habermas, J. (1995, March). Reconciliation through the public use of reason: Remarks on John Rawls's Political Liberalism. In Journal of Philosophy. xcii, (3) pp. 100- 131.

Jin, K.K. (1991). *Whose war? What peace? Reflections on the Gulf Conflict.* Aliran Kesedaran, Malaysia: Penang.

Jones, S. (1995).*Cybersociety: Computer-mediated communication and community*. London: Sage.

Kazan, E.F. (1993). *Mass media, modernity, and development: Arab states of the Gulf.* Westport, CT: Praeger.

Kiathe, R. (1 995, Winter) BITNIS in Nairobi. In *Project for Research Libraries. 4*, (21) p. 6.

Kisiedu, C. (1994, Winter) The West African Library Association: Its origins, break- up and revival. In *Project for African Research Librarians, 4,* (2), pp. 1-6.

Land, M. (1994). Computers empower independent newspapers in Cameroon. In *African Technology Forum. 7*, 2.

Lawrence, S. & Lief, L. (1994, March). The China syndrome. In *U.S. News and World Report. 2,* 1, pp. 39-42.

Lee, D. (1986, May) *Diffusion of imperialism: A view of communication, culture, mass media and technology*. Paper presented at the 35th. Annual Conference of the International Communication Association, Honolulu, Hawaii.

Lindhorst, K. (1991). New telecommunication trends and international relations. In H. Mowlana and N. Levinson. (Eds.) *Telecommunications and international relations: An East-West perspective.* Washington, D.C. : The American University, International Communication Program.

Lomax, D. (1995). Financing the telecommunications sector in sub Sahara Africa. In *Africa Communications. 6*, (1) pp. 14-15.

Luostorinen, H. (1996, October). Innovations of Moral Policy in the Gulf War. In *Media Development.*

Mann, K. & Roberts, R. (Eds.). (1991). *Law in colonial Africa.* Portsmouth, NH: Heinemann.

McGraw, D. (1996, Oct 28). Happily Ever Nafta? In *US News and World Report, 21*, 17.

Moo, J. (1981). The ASEAN computer scene-The Problems Ahead. In J.M. Bennett & R.E. Kalman (Eds.) *Computers in developing countries.* NY: North-Holland.

Mowlana, H. (1996). *Global communication in transition: The end of diversity?* Thousand Oaks, CA: Sage.

Mowlana, H. (1994). From Technology to Culture. In G. Gerbner, H. Mowlana, & K. Nordenstreng (Eds.). *The global media debate.* (pp. 61-66). Norwood, NJ: Ablex.

Mowlana, H. (1994). Shapes of the future: International Communication in the Twenty First Century. In *Journal of International Communication*, 1, 1.

Mowlana, H. & Kamalipour, Y. (eds.). (1994). *Mass media in the Middle East: A comprehensive handbook.* Westport, CT: Greenwood Press.

Mowlana, H. & Wilson, L. (1990). *The passing of modernity. Communication and the transformation of society.* NY: Longman.

Mowlana, H. (1986). *Global information and world communication: New frontiers in international relations.* White Plains, NY.: Longman.

Mowlana, H. (1975). The Multinational Corporation and the Diffusion of Technology. In A.A. Said & L.R.S. Simmons (Eds.) *The new sovereigns: Multinational corporation as world power.* Englewood Cliffs, NJ: Prentice Hall.

Newsome, D. & Wolbert, A.J. (1988). *Media writing: Preparing information for the mass media.* Belmont, CA: Wadsworth Publishing.

Ngwainmbi, E. (1999). Black Connections & Disconnections in the Global Information Supermarket. (Chapter 9). In J.T. Barber & T.T. Tait (Eds.), *The information society and the Black community.* Connceticut:Greenwood Press.

Ngwainmbi, E. (1998). New Nation in Cyberspace. In *AfricaAccess Magazine, (1)*, 3 p. 9.

Ngwainmbi, E. (1997). Technology and Information Flow in Africa. In *Economic Development Institute Forum*, The World Bank, 2, pp8-9.

Ngwainmbi, E. (1995a). *Communication efficiency and rural development in Africa*. New York, NY.: University Press of America.

Ngwainmbi, E. (1995b). Indigenous Information Sources in a Traditional African Society. In C.Okigbo (Ed.).*Media and sustainable development* (pp.389-408). Nairobi, Kenya: Media Congress.

Ngwainmbi, E. (1991). *The effectiveness of communication in rural development in Cameroon: A case study of two communities*. Dissertation. Washington, D.C.: Howard University.

Nwabuzor, E. (1980). Ethnic Value Distance in Cameroon: In J.N. Paden, (Ed.), *Values, identities, and national integration*. Evanston, IL: Northwestern University Press.

Nyang, S (1995, April) The Road to Nationhood: The Political Thought of Amilcar Cabral. In *New Directions*.

Odumosu, J.0. (1993, June). After 10 years: The Future of West African Librarianship. In *Library World. 64*, (756), pp. 358-359.

Okuwoga, 0. (1990). Impact of Information Technology on Nigeria's Socio-economic Development. In S.C. Bhatnager & N. Bjorn-Anderson (Eds.) *Information technology in developing countries*. North Holland: Elsevier Science Publishers.

Orondo, P. (1994). Internetworking with TCP/IP. In *African Technology Review.* 7, 2.

Patton, P. (1986). *Open road*. NY.: Simon & Schuster.

Pizarro, O.F. (1994, Sept.) NAFTA: Bad News for Mexico? In *World Press Review,* 41, (9) 40.

Pratt, C. (1995, August) *New communication technologies for development in Africa: A strategic issues-management approach.* Research paper presented at the 78th Annual AEJMC Convention, Washington, D.C.

Radway, R. (1983). Anti-Trust Technology Transfers & Joint Ventures in Latin American Development. In *Lawyer of the Americas*, 15.

Reingold, H. (1993). A slice of life in my virtual community. In L.M. Harasin (Ed.) *Global networks* (pp. 57-80). Cambridge, MA.: MIT Press.

Roffe, P. (1985). Transfer of Technology: UNCTAD's Draft International Code of Conduct. In *The International Lawyer,* (19), 2.

Root, F. (1968, Summer). The Role of International Business in the Diffusion of Technological Innovation. In *Business and Economic Bulletin*. 24, pp. 19-27.

Schiller, H. (1970). Transnational Media and National Development. In K. Nordenstreng 7 H. Schiller (Eds.). *National sovereignty & international communication: A reader.* Norwood, NJ: Ablex.

Seel, B. P. (1995, August). *An Alternative Model for the Study of International Telecommunication Regulation: A Case Study of HDTV in Japan, US, and the European Union.* Paper presented at the 78th Annual Convention of AEJMC,Washington, D.C.

Seligman, A. (1992). *The idea of civil society*. NY: The Free Press.

Seyoum, B. (1990). *Technology licensing in East Africa*. London, England: Averbury.

Sikes, A.C. (1995). Policymaking Regarding Citizen Information Services. In R. Williams & J. Pavlik (Eds.), *The people's right to know: Media, democracy, & the information highway*. Hillsdale, NJ: Lawrence Erlbaum Associates, Inc.

Singer, H. (1977). *Technologies for basic needs*. Geneva: International Labour Organisation.

Stewart, F. (I 97 7). *Technology and underdevelopment*. London: Macmillan.

Stover, J.W. (1984). *Information technology in the Third World*. Boulder, CO.: Westview Press.

Symonds, W. (1992, June 29). Is Canada ready for the next MCI?*Business Week*. p. 36.

The Application of Computer Technology for development (1971). United Nations, New York: Department of Economics and Social Affairs.

Truxall, G.J. (1990). *The age of electronic messages*. NY.: McGraw Hill.

Turow, J. (1992). *Media systems in society*. NY.: Longmans.

Vedel, T. (1996). French Policy for Information Superhighways: The End of High-Tech Colbertism. In *Information Infrastructure & Policy, 5,* (1), pp. 41-69.

Weber, M. (1976). *Wirtschaft und Gesellschaft*. Tubingen, The Netherlands

Westendoerpf, D. & Odeh, A. (1990). The Case for Investing in Telecommunications in Africa. In *Telecommunication Proceedings of Africa Telecom 90: December 3-6*. Harare, Zimbabwe.

Whalen, P. (1995, August). *Mobile satellite communications: From obscurity to overkill*. Research paper presented at the 18th AEJMC Convention,Washington, D.C.

Wilson, L.R. S. (1995). *Mass Media, Mass Culture*. NY.: McGraw Hill.

Woherem, E. (1993). *Information technology in Africa*. Maastricht,The Netherlands: ACTS Press.

Work, C. (1990, Feb. 26). *US News and World Report*. p. 44.

World Bank (1993a). *The World Bank Report, 1993*. Washington, D.C.

Zeigler, B. (1993, March 8). AT&T Reaches Way Out For This One. *Business Week*.

Appendix I

Questionnaire for African Embassy Staff

This questionnaire is designed to evaluate the views of African Embassy workers in Europe & America regarding the exportation to and use of telecommunication equipment in Africa. Please check the response(s) you consider most appropriate. Do not write your name.

1. What is your profession?

2. Circle all of the systems below which you own.

   a. TV set
   b. VCR (video cassette recorder)
   c. Telephone
   d. Cordless phone
   e. Computer
   f. Fax machine
   g. Satellite dish
   h. Beeper
   i. None of the above

3. I use the above system for:
   a. Business only
   b. Entertainment only
   c. Transmitting information
   d. Both a & b
   e. All of the above

From question 4 to 11, please circle only one response

4. a. I have sent or I will send the systems to Africa
   b. I am using or I will use them here

5. I would like to see ———————— systems shipped to Africa

   a. More
   b. None
   c. Some (name them)

6. If you answered b or c give your reason(s). Use additional paper, if necessary

7. Members of the global information supermarket pay a fee in order to have access to business, social, and economic information. Do you agree that your country should become a part of that supermarket?

   a. I strongly agree
   b. I agree
   c. I scarcely agree
   d. I disagree
   e. I strongly disagree

8. Are you concerned that your country might lose its autonomy to more influential countries if it uses their telecommunication equipment and services?

   a. I am very concerned
   b. I am somewhat concerned
   c. I am less concerned
   d. I am not concerned

9. For how long have you owned any of the systems you mentioned in question 2?

a. 1 to 6 months
b. 7 months to 1 year
c. 2 to 5 years
d. 6 years plus

10. Would you prefer an Africa-based and African-owned industry to manufacture and sell telecommunication equipment?

a. Yes
b. No

Give reasons for your response.

11. How long have you lived abroad?

a. 1 to 6 months
b. 7 months to 1 year
c. 2 to 5 years
d. 6 years plus

Appendix II

Questions for the African Public

Please circle the response you consider most appropriate.

1. I am:

    a. An educated person
    b. A business person
    c. Both
    d. None of the above

2. Do you do business abroad?

    a. Yes
    b. No

3. If you answered *a* by what means do you contact your partners?

    a. Facsimile
    b. Cordless phone
    c. Telephone

d. Writing letters
e. Making trips abroad
f. Other means. Please specify

4. If you use the telephone or facsimile, approximately what is your call completion rate? In other words, how many times do you dial before the line goes through?

a. Once
b. Twice
c. Three times
d. Four times
e. Five times and more

5. Which one of them goes faster?

a. Telephone
b. Facsimile

6. Which types of information products do you import? Circle all that apply.

a. Fax machines
b. Radio sets
c. elevision sets
d. Beepers
e. Satellite dishes
f. Cordless phones
g. Computers
h. Telephones

7. How old are you?

a. 16-21 yrs.
b. 22-29 yrs.
c. 30-35 yrs.
d. 36-49 yrs.
e. 50 yrs. or older

Appendix III

A One-Week Seminar Program

Training Communication Experts in the Ministries Education and Culture on Primary Education Campaigns

Day One:	**COMMUNICATION AND SOCIOECONOMIC DEVELOPMENT**
Session I:	-Define the role of education in socioeconomic development *Break* -Identify types of societies in West and Central Africa *Lunch*
Session II:	-Describe the roles of communication in development -Determine the relationship between communication, the communication expert (practitioner), the law maker (Minister), and media campaigns -Quiz

Day Two: ROLE OF COMMUNICATION IN PROMOTING EDUCATION

Session I: -Dedicating resources toward a primary education (PE) campaign
Break
-Review strategies for developing education campaigns
Lunch
Session II: -Ways of effectively carrying out PE campaigns
-Questionnaire and Quiz

Day Three: TOOLS AND MEANS

Session I: -Identify and describe communication/campaign materials (participatory video, public gathering, church service, music, theatre, poster, newspaper, radio, television, ritual)
Break
-Review applicable materials and discuss their usefulness in promoting PE ideas and messages
Lunch
Session II: -Conduct a workshop on the techniques of using effective communication systems

Day Four: PARTICIPANT OBSERVATION: FEEDBACK

Session I: -Discuss participant's field experiences
Break
-Watch and discuss videos on campaign strategies. focus on participatory videos
Lunch
Session II: -Identify target groups for the campaign
-How to use interpersonal models to approach potential pupils in rural, impoverished, and inner city communities

-New ways of using radio, TV, newspaper time, and space
-Quiz

Day Five: TARGETING TRADITIONAL AND GOVERNMENT AUTHORITIES

Session I: How to use teachers, parents, traditional leaders, healers, and local authorities as information outlets
Break
-How to handle traditional leaders' sensitivities toward providing education information to youngsters
Lunch

Session II: How to manage a Minister's insensitivity regarding communication as a development agent
-Questionnaire

Day Six: MEDIATION AND NEGOTIATION OF POWER

Session I: -How to persuade Ministers to invest in communication as a campaign projects, e.g. allocate campaign funds
Break
-How to manage the Minister's time and follow-up meetings
Lunch

Session II: How to use interpersonal models to help the potential pupil negotiate his/her "social (free) time" and allocate more time for "formal" education
-Questionnaire
-Seminar evaluation

Author Index

A

Adekola, S.A., 133, 217
Akwule, R., 177, 217

B

Barrym, M., 154, 217
Becker, J., 93, 217, 218

C

Cabanellas, G., 157, 218
Cappel, J., 38, 218
Copeland, A., 26, 27, 218

D

Davidson, O., 73, 128, 218
Davies, M., 104, 218
DiSalvo, D., 117, 218
Dordick, S., 112, 218

J

Jones, S., 219

K

Kamalipour, Y. 170, 220
Kazan, F. 170, 219
Kiathe, R. 125, 219
Kisiedu, C. 124, 219

L

Lee, D., 17, 28, 36, 219
Lindhorst, K., 35, 219
Lomax, D., 145, 219

M

Mowlana, H., 16, 35, 36, 114, 170, 219, 220

N

Ngwainmbi, E., 20, 76, 99, 100, 132, 134, 202, 220, 221
Nwabuzor, E., 73, 76, 221
Nyang, S. 96, 221

O

Odeh, A. 148, 222
Okuwoga, O. 33, 221
Orondo, P., 18, 221

P

Patton, P., 107, 221
Pizarro, F., 157, 221

R

Reingold, H., 108, 221
Roberts, R., 157, 220
Roffe, P., 39, 221

S

Schiller, H., 109, 221
Seyoum, B., 40, 222
Sherman, L., 26, 27, 218
Singer, H., 133, 222
Stover, W., 115, 222

V

Vedel, T., 57, 222

W

Weber, M., 93, 222
Whalen, P., 56, 222
Wilson, L., 16, 21, 220, 222

Z

Zeigler, B., 222

Subject Index

A

ability to manage, 74
academic, 81, 199, 206, 215
administrators,
73, 97, 134, 148, 183
advanced societies, 107, 114
African colonies, 124
African countries, 73, 89, 94,
105, 115, 116, 133, 137,
139, 140, 144, 145, 146, 147,
148, 150, 151, 193
African culture, 195, 203
African governments, 74, 75, 76
101, 126, 144, 145, 147
African public, 75, 76, 77, 81
African societies, 73,
77, 97, 118
Africentric, 73
Africentric scholarship, 73
Afrikaans, 173
age of enlightenment, 118
aggressive behavior, 175
agricultural centers, 96
ALCATEL, 143, 149
American university, 120, 197
amounts of, 105,
analyzing foreign competition,
137
authorities should, 129
autonomy,
75, 82, 83, 143, 150
availability, 24, 74, 85, 101, 112,
128, 137, 171, 198, 206

B

backbone of the economy, 136
bad economy, 100
beneficiary, 17
Benin, 89, 150
BITNET, 125
Blacksmiths, 134
blacksmiths, 135, 136
Brazil, 27, 128, 138, 142
bribe-givers, 116
Business,
71, 116, 142, 176, 185
business opportunities,
195, 208, 215
business respondents, 78, 79

by 2000AD, 149, 171, 175

C

Cameroon, 32-33, 76, 89, 92,
106
106, 115, 124, 134, 141, 144,
career 143, 213
cash machines, 103
CD-ROM, 26, 30, 105, 124
Cecil Rhodes, 91
cellular phone, 151
CFA, 141, 145, 147

Chamber of Commerce,
139, 140, 142
Chile, 142
colonial administrators, 134
colonial capitalism, 137
Congo, 150
connection cost, 144
consumer,
cyberspace, 60, 165, 131-132,
140, 141, 146, 151

convenient sales, 142
cosmologies, 95
cronies, 147, 156
cultural context, 154
cyberspace, 60, 165

D

decision-making, 72, 83
demand for, 78, 79, 81, 171
deregulation, 58, 65, 131
deterred progressive change, 118
development information,
108, 121,129, 200, 201, 204, 206, 207
dialogue,
66, 92, 99, 106, 149, 165, 179, 214
diffusion, 132
digital information, 58
dissemination of, 124, 183, 207
distance learning, 113, 148, 192
dysfunctional, 100

E

ecological systems, 95
economic bankruptcy, 137
economic significance, 114
economic status, 61, 136
ecosystem, 127, 198
effective telephone service, 49
embracing foreign technology, 94
enhance the development of, 164
environmental marketing, 204
ethnic background, 153
European, 49-50, 65, 135, 147, 149,
156, 157, 170, 180
European countries, 50, 65, 149
evangelists in Africa, 133
exportable product, 142
exportation cost, 132
exportation problems, 142

F

facsimile,
78, 93, 103, 149, 151, 163, 171
fast rate, 131
FBI, 154
FCC, 54, 58
fiber optic, 47-49, 60, 63, 67-69
75, 127
flute, 106
foreign assistance, 82, 105
foreign goods, 93, 102
foreign governments, 72
foreign investment, 115
foreign markets, 67
France, 56, 57, 61, 115
Francophone nations, 115
free market economy, 85
French-based, 115
fuel consumption, 128

G

garner profit, 143
Germany, 56, 61, 139
global communication, 137
global information marketplace,
137
global interaction, 161
global reality, 82
government policies, 98, 143,145
government representative, 75
growth potential, 48
Guinea, 89, 96, 163, 169

H

health information, 106
holistic, 99, 101, 104, 118
human capacity, 99

I

ideology, 111
imminent military invasion, 133
impact of, 83, 89
imperialism, 98, 111
implementing policies, 161, 165
indigenous culture, 110, 194
individual living standards, 97, 98
industrial revolution, 98, 107
inferior feeling, 114
information economy, 103, 124
information products, 102, 115
information resources, 120, 104
information society, 92, 173
information superhighway, 170, 181
installation of, 170, 187
internet users, 95, 112,113, 126, 191
IT markets, 85
Ivan Sertima, 73
Ivory Coast, 144, 150

J

jingling bells, 134, 136

K

Kenya, 89, 121, 170, 171
kinsfolk, 97, 100, 102

L

lack of, 17, 19, 87, 109, 118, 120, 182, 193
landlocked countries have, 210
Lesotho, 140, 163, 172, 173
local interpreters, 60
local technology, 136, 137
locally-oriented business, 81
louder sounds, 134
low importation, 81

M

Mali, 150
manual labor, 134
market capabilities, 71
market shares, 54, 55
market size, 52, 61
maximization, 75, 82, 160
MCI, 47, 51-52, 55, 58, 63-64, 144
media campaign, 68, 92, 184, 200
media experts, 181
media practitioner, 113
media practitioner, 20, 165
medical examinations, 128
message transmission, 125, 208
Mexico, 26, 46-48, 116, 157, 158
Ministry of Finance in Morocco, 105
musical instrument, 135
Muslim, 26, 73, 169, 170

73, 92, 93, 111, 127, 133, 134,
mobile radio, 46, 201
monarchs of, 117
Motorola, 46, 47, 48
musical instrument, 135
Muslim, 26, 73, 169, 170

N

NAFTA, 47, 48, 116, 157
national and international experts, 175
Nelson Mandela, 139
neo colonial, 133
new markets, 57, 89, 110, 115-116, 160
NGOnet, 125
NWICO, 47, 120
NWIO, 87

O

obtain new information, 125, 190
ownership, 58, 68, 81-82, 86, 145, 174, 176

P

PANAFTEL, 68, 149, 150
participatory video, 184, 203, 204
partners abroad, 78
patriotic, 75, 111, 124
PC computers, 121
peace corps volunteers, 127
personal property, 143
political candidates, 113
political climate, 40, 85, 98, 166, 177
political crises, 70, 72, 73, 128, 146
political history, 72
political image of, 72
post-modern technology, 32, 136
potential benefits, 104
potential investors, 103
poverty rate, 105, 129
power keepers, 98
preserving information, 16, 148
private companies, 144, 159
privatization, 160-161, 163-164, 177-178, 179, 190, 207
privatizing the, 160, 163, 165, 177, 178
product ownership, 82, 86
progressive communication, 17
property rights, 38, 97, 173
provide better services, 57
providing information about, 72
public gatherings, 13, 15, 108, 129, 184
public records, 105, 164
public relations firms, 30, 93
purchasing power, 74, 89, 132, 166

R

Radio France Internationale, 168
raising techniques, 130
rapid development, 114, 173
RASCOM, 32, 68, 149
receptionists, 83, 85, 86, 88
record-keeping , 164, 180
regional information, 104
regular consumers, 92

regulatory framework, 54, 145, 157
religious organizations, 205
researchers in Africa, 105
residential areas, 51
respondent traits, 74
respondents, 74, 76, 78-84, 85, 88-89
responsibilities of, 68
retailers, 65, 67, 102,207

S

sacred place, 127
sales agencies, 82
sales people, 102
Salmon Rushdie 169
satellite dishes, 11, 20-21, 27, 35 78-79, 80, 82, 84, 86, 93
Scotland Yard, 99
self empowerment, 70
sending data, 125
Senegal, 61, 89, 97, 103, 115, 144 , 150, 169, 170
Singapore, 138, 142
socialization patterns, 93
socioeconomic paradigms, 99
software, 19, 28, 30, 32-34, 40-41, 104, 122, 125, 210, 212
Sotho languages, 173
South Africa, 89, 91, 121, 129, 139, 140
spare parts, 133
stable economy, 75, 101, 168
standard of, 96, 97, 171
strategic investors, 144
strategies for, 102, 163, 180, 181
strategies to entice, 52
students in, 119, 120
sub-groups, 109
sub-Saharan, 94-95, 124, 128-128, 129, 140, 145, 160, 202
Sudan, 89
sufficient profit, 143
Supreme Force, 118
sustainability, 122, 185
sustainable operation, 141

T

talking drum, 133
Tanzania, 89, 92, 194, 210
teachers, 82, 83, 183, 184, 200, 205
technical capabilities, 113
technology consumers, 157
technology for, 94, 136, 189,207
technology transformed, 96
telecommunication networks, 57
telegraph wires, 91
telephone callers, 79
telephone companies, 51
telephone dial, 76
telephone exchange, 115, 149, 171
telephone service, 49, 54, 57, 63, 103, 140, 144, 148
telephone transactions, 63, 126
Thailand, 142
210
The African mind, 98, 99
the creation of, 50, 132, 148, 177,
the Ministries of, 181, 182, 186,

the Gulf War, 140, 169
the marketing of, 135, 159, 205
the Middle East,
50, 61, 73, 169, 170
The Ministries of, 181, 182, 186, 208, 210
the Soviet Union, 70
the spread of, 192
the telephone as, 52, 59
The World Bank, 141, 144, 147, 179, 180, 185,193, 215
the year 2000,
63, 144, 169, 171, 193
theories of, 96
Third World markets, 45, 131
toward modern telecommunication technology, 59
traditional Africans, 117, 136
traditional rulers, 77, 157
traffic flows, 121
traffic management techniques, 161
transferring technology, 158
transportation costs, 141
transportation of, 114
tribal leaders, 106
tropical climate 168

U

UNDP, 141
unresearched, 119
UNZANET, 121, 210
up-to-date information, 122
urban communities,
184, 197, 198, 206
use of telecommunication equipment , 49, 74, 141
using beepers, 79

V

value of, 81, 85, 135, 189-190, 203 214, 215
viable technologies, 133
village residents,
76, 127, 133, 134
virtual community, 126
Voma, 134
VSAT, 138, 180

W

West African Journal, 124
West African Library Association, 123
widespread use of, 85
World Wide Web, 192
Worldwide Digital Fax, 151

Z

Zaire, 140, 144
Zambia, 91, 121, 166, 177-178, 192
Zambia Privatization Agency, 177
ZAMTEL, 68

About the Author

Dr. Emmanuel Kombem Ngwainmbi is a specialist in International Strategic Communication and Third World Policy & Development. His research interests include development communication and telecommunication management.

He is a communications consultant for the United Nations, World Bank, IMF, US Departmentof State's Foreign Service Institute, and other major international organizations.

Dr Ngwainmbi has done comprehensive telecommunication policy and economic development studies for Chad, Senegal, South Africa, Mexico, Botswana, Peru, Kenya, Cameroon, and Zambia.

He is also an Associate Professor of Communication, who has lectured in numerous colleges and universities in the US and abroad, including the National Center for Communication Studies, George Washington University.

He holds Ph.D,MA, and BA degrees from Howard, Jackson State, and Yaounde universities, respectively.

He is a journalist, the Technology & Development Editor for *Africa Access Magazine*, member of the editorial board of several Africa-based magazines and of several international professional organizations. He has authored numerous scholarly articles and books, including *Communication Efficiency & Rural development in Africa,Dawn in Rage,* and *A Bush of Voices.*